Sustainability Metrics and Indicators of Environmental Impact

Sustainability Metrics and Indicators of Environmental Impact

Industrial and Agricultural Life Cycle Assessment

Eduardo Jacob-Lopes
Federal University of Santa Maria, Santa Maria, Rio Grande do Sul, Brazil

Leila Queiroz Zepka
Federal University of Santa Maria, Santa Maria, Rio Grande do Sul, Brazil

Mariany Costa Deprá
Federal University of Santa Maria, Santa Maria, Rio Grande do Sul, Brazil

ELSEVIER

Elsevier
Radarweg 29, PO Box 211, 1000 AE Amsterdam, Netherlands
The Boulevard, Langford Lane, Kidlington, Oxford OX5 1GB, United Kingdom
50 Hampshire Street, 5th Floor, Cambridge, MA 02139, United States

Notices

Knowledge and best practice in this field are constantly changing. As new research and experience broaden our understanding, changes in research methods, professional practices, or medical treatment may become necessary.

Practitioners and researchers must always rely on their own experience and knowledge in evaluating and using any information, methods, compounds, or experiments described herein. In using such information or methods they should be mindful of their own safety and the safety of others, including parties for whom they have a professional responsibility.

To the fullest extent of the law, neither the Publisher nor the authors, contributors, or editors, assume any liability for any injury and/or damage to persons or property as a matter of products liability, negligence or otherwise, or from any use or operation of any methods, products, instructions, or ideas contained in the material herein.

Library of Congress Cataloging-in-Publication Data
A catalog record for this book is available from the Library of Congress

British Library Cataloguing-in-Publication Data
A catalogue record for this book is available from the British Library

ISBN: 978-0-12-823411-2

Publisher: Charlotte Kent
Acquisitions Editor: Marisa LaFleur
Editorial Project Manager: Sara Valentino
Production Project Manager: Sruthi Satheesh
Cover Designer: Victoria Pearson

Typeset by SPi Global, India

Contents

Authors' biographies ix
Foreword xi

1. **Sustainability metrics and indicators through the life cycle assessment: A brief history**

2. **Fundamentals of life cycle assessment: Definitions, terminology, and concepts**

Phases for life cycle assessment 8
Objective and scope 8
Life cycle inventory 13
Life cycle impact assessment 18
Interpretation 22
Boundaries for life cycle assessment 23
References and recommended reading 26

3. **Methods of evaluation of the environmental impact on the life cycle**

Impact categories 31
Ozone depletion potential (stratospheric) 31
Photochemical ozone creation potential (tropospheric) 33
Ecotoxicity 35
Acidification potential 38
Eutrophication potential 40
Energy assessments 43
Water footprint 52
Land use 56
Global warming potential 59
Human toxicity potential 63
Categories of damage or areas of protection (endpoints) 63
References and recommended reading 67

4. Optional elements of life cycle impact assessment

Grouping 71
Normalization 71
Weighting 74
References and recommended reading 76

5. Assistant's tools toward life cycle assessment

Life cycle thinking 77
Social life cycle assessment 79
Life cycle costing 83
Life cycle sustainability assessment 86
Defining the problem, generating alternatives, and establishing criteria 87
Assigning criteria weights 87
Construction of the evaluation matrix 89
Sensitivity analysis 90
References and recommended reading 90

6. Carbon footprint and carbon market

Carbon footprint assessment standards 91
Selection of greenhouse gases 94
The methodology procedure of evaluation of carbon footprint 95
Methodological applications 96
System delimitations 98
Mathematical modeling 100
Carbon certifications 102
Market carbon 104
Carbon credits 106
Policy strategies among carbon footprint, carbon market, and carbon pricing 108
Carbon market trades 111
References and recommended reading 115

7. The role of life cycle assessment in agricultural systems

Functional unit for agricultural systems 118
System boundaries 119
Allocation 122
Quantification bottlenecks in the life cycle inventory stage of agricultural systems 125
Land use 128
Fertilization of agricultural crops 133
Other parameters for agricultural systems 137
References and recommended reading 140

8. Economic aspects of sustainability

An approach to the economic aspects of corporate institutional sustainability 146
Strategies for the manifestation of the economic aspects of institutional sustainability 148
An approach to the economic aspects of sustainability from the perspective of the shared economy 151
An approach to the economic aspects of sustainability through sustainable added value 155
Absolute measures 155
Relative measures 156
Linking relative and absolute measures 157
Guidelines and contextualization of changes in the concepts of efficiency and effectiveness 160
Opportunity costs: Concepts and application importance 162
References and recommended reading 169

9. Sustainability metrics and indicator strategies for a modern society

Sustainability tripod dimensions 175
Environmental dimension 175
Economic dimension 175
Social dimension 176
Suggestions for the development of sustainability metrics and indicators 177
Beyond sustainability 180
References and recommended reading 182

Annex 183
Index 197

Authors' biographies

Dr. Prof. Eduardo Jacob-Lopes is currently associate professor at the Department of Food Technology and Science, Federal University of Santa Maria. He has more than 15 years of teaching and research experience. He is a technical and scientific consultant to several companies, agencies, and scientific journals. He has 600 publications/communications, which include 15 books, 60 book chapters, 120 original research papers, 400 research communications in international and national conferences, and has registered 15 patents. His research interest includes biotechnology and bioengineering with emphasis on microalgal biotechnology.

Dr. Prof. Leila Queiroz Zepka is currently an associate professor at the Department of Food Technology and Science, Federal University of Santa Maria. She has more than 15 years of teaching and research experience. She is a technical and scientific consultant to several companies, agencies, and scientific journals. She has 480 publications/communications which include 15 books, 60 book chapters, 100 original research papers, 300 research communications in international and national conferences and has registered 11 patents. Her research interest includes microalgal biotechnology with emphasis on biomolecules and natural pigments.

Dr. Mariany Costa Deprá is currently a researcher at the Department of Food Technology and Science, Federal University of Santa Maria. She has a masters and doctorate in food science and technology from the Federal University of Santa Maria. She has published more than 50 publications/communications, including 17 original research papers, 12 book chapters, 30 research communications at national and international conferences, and has registered 1 patent. She has experience in the field of environmental biotechnology, with an emphasis on microalgae and mass balance. Her research interest includes process engineering with an emphasis on sustainability metrics and indicators and life cycle assessment.

Foreword

Will exclusive economic criteria continue to lead the way for the industrial and agricultural development of the future? Indeed, some countries and companies are considering their production platforms based only on economic aspects.

However, sustainable development moves a step further, thinks ahead, and deals wisely with the equilibrium of the resources. Sustainability is not hype; it is needful.

The term "sustainability" or "sustainable development" was defined for the first time in 1987 in the Brundtland Report Our Common Future. Since then, the world is a close look at this concept.

Nevertheless, in the last 40 years, little concrete progress has been achieved based on this idea. In fact, the economic priorities associated with the lack of consistent metrics and indicators to evaluate sustainability have delayed its effective application.

The sustainability metrics and indicators of processes and products are part of this consolidation, being, nowadays, the life cycle assessment the most consistent tool for the transition of the theory to the practice in sustainable development.

This specialized book is intended to be a guideline for the application of life cycle assessment. The chapters are organized sequentially to provide the progression of the phases of the sustainability tools consistently so they can be used as a textbook by undergraduate and graduate students as well as environmental science professionals.

Therefore, Chapter 1 briefly presents the history of the first environmental tool, the life cycle assessment. Thus, through a timeline, this section approaches the main events leading to the emergence of this assessment instrument, as well as the first step toward its standardization based on the sustainable society-economy transition.

Chapter 2 shows in detail the fundamentals of life cycle assessment. Thus, after a slight introduction, the reader is led to know all the including phases of a complete analysis. In addition, this guide presents some alternatives and boundaries for the use of this environmental tool.

Chapter 3 consists of enabling the reader to deepen knowledge about the methods of assessing life cycle impacts. Furthermore, this chapter allows undergraduate students as well as professionals who make use of this tool to standardize the quantification of environmental impacts through mathematical models established by the impact categories already consolidated.

Chapter 4 provides a major understanding of the optional elements of life cycle assessment. This chapter requires broader attention since the data obtained from impact categories will be compared to external data in the scientific literature.

Chapter 5 aims to contribute theoretically to the state of the art of life cycle assessment tools. This chapter presents new systemic strategies, which aim to identify and improve the gaps associated with social and sustainable processes. Thus, the life cycle assessment such as life cycle thinking, life cycle cost, social life cycle assessment, and life cycle sustainability assessment are discussed in this section to expand the frontiers of knowledge and to outline its consolidations in the future.

Chapter 6 provides an overview of the relationship of carbon footprint to sustainability metrics and indicators and its influence on environmental impacts. It also addresses the perspectives of the global carbon market and the main carbon credit quantification methodologies.

Chapter 7 consists of a critical analysis of what is already done, and what can be done in the future regarding sustainability metrics for the agricultural systems, in addition to addressing the key impact assessment indicators of these segments.

Chapter 8 aims to provide a comprehensive approach to the economic aspects associated with sustainability indicators. In addition, the efforts made over the years to develop a systematic vision are presented, which results in a cultural change in the face of the paradigms that involve economic and environmentally sustainable production systems.

Finally, Chapter 9 consists of a critical analysis of the establishment of sustainability metrics and indicators. Besides, it provides a continuous approach involving complex social, environmental, and economic factor assessments. Still, it presents slight bottlenecks that persist throughout the evolutionary cycle of this tool, considering criteria beyond sustainability, that to the moment, need to be considered. In the closure, it approaches the final background and future perspectives for the success of the sustainability metrics and indicators.

Chapter 1

Sustainability metrics and indicators through the life cycle assessment: A brief history

Population growth and the consequent intensification of industrial activities have sharply boosted the many environmental problems that exist in the world, including the extraction of natural resources and climate change. In view of this, political-governmental pressures were imposed on the industrial sectors, with the aim of improving environmental performance. Thus, definitions and concepts about sustainability have emerged, becoming a major apparent demand in the management of environmental systems.

Initially, at the beginning of the 20th century, environmental indicators were not part of the quantification of industrial processes. Therefore, the inputs and outputs of the systems were basically applied to the industrial sectors aiming the economy. However, once monetary requisition of the projects was demanded, reports were published with the first quantifications of energy inputs and outputs. These publications turned to a greater understanding of the interdependence of sectors, focused on a circular system model (Leontief, 1936).

Over time, with the industrial revolution in the mid-1960s, a series of agreements, conventions, and laws emerged with an interest in minimizing the effects of human activities on the environment. Faced with these challenges, it was important to assess the real environmental impact of economic development. Thus, preliminary research such as the Leontief focused exclusively on energy and resource analyzes (Smith, 1963).

Resource and environmental profile analysis (REPA) was developed to follow the production chain in an attempt to have a broad view of the life cycle of a given product (Hunt, Franklin, & Hunt, 1996). Thus, the application of this tool in the production of waste and in the use of energy was increasingly used by the private sector to better understand its production systems, insofar as the interest of public agencies also existed in the 1970s.

Sustainability Metrics and Indicators of Environmental Impact. https://doi.org/10.1016/B978-0-12-823411-2.00001-3

Among the leading private sectors in the use of this tool is the beverage industries. In 1969, a multinational beverage company commissioned the first study to examine the full environmental impact of a package, setting the framework for the life cycle assessment methodology used today (Guinee et al., 2010). The focus on life cycle management used by the company has enabled substantial advances in the use of sustainable recyclable materials. This ensures an improvement in production efficiency and effectiveness in the application of life cycle assessment.

Given this scenario, in the early 1990s, with the end of the confidentiality of private reports, and the rise of the development of the society for environmental toxicology and chemistry (SETAC) shaped the development of what is now known as life cycle assessment (LCA) (Fig. 1.1). Therefore, to standardize the stages of LCA application, the basic structure of LCA developed by SETAC presented the following arrangement: definition of objective and scope, inventory assessment, impact assessment, and interpretation (Fig. 1.2).

Nevertheless, after standardization of the initial methodology, advances were presented in front of the series of ISO 14000 standards. Thus, the evolutionary of the life cycle assessment presented well-established steps in relation to the general principles, the targeting for the definition phases of objective and scope, inventory assessment, and how to compute the possible environmental data associated with human and industrial activities. In addition, the way in which the results were interpreted and the rules on how to present the data to the interested public were developed exclusively (Fig. 1.3).

In this way, the chapters present in this book are intended to help provide a deeper understanding of challenges for the application of life cycle assessment tool aiming to establish the sustainability metrics and indicators. Thus, the main purpose of this book is to enable the reader to develop a clear and deep understanding of the fundamentals in the analysis of processes and products in the application life cycle assessment in industrial processes and agricultural systems. In addition, the chapters in this book are intended to contribute substantially to critical thinking in the application of this tool. Therefore, these objectives will be achieved through the presentation of a structural methodological basis of the life cycle assessment, which will allow the reader to continuously apply these approaches in a wide variety of problems of sustainable evaluation of the industrial processes and agricultural systems.

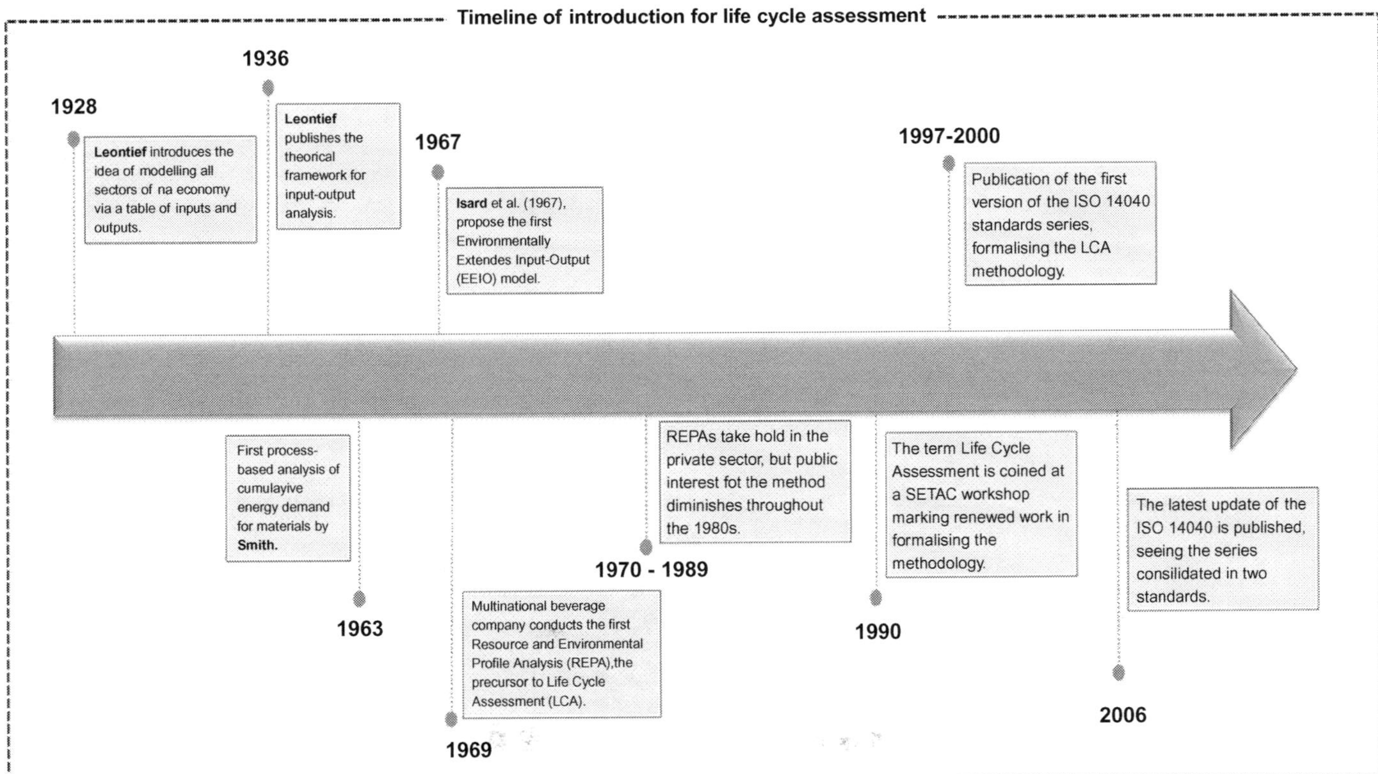

FIG. 1.1 Timeline of the emergence of the life cycle assessment.
This figure addresses the evolutionary timeline of the life cycle analysis tool. *(Reproduced with permission from Crawford, Bontinck, Stephan, Wiedmann, and Yu (2018).)*

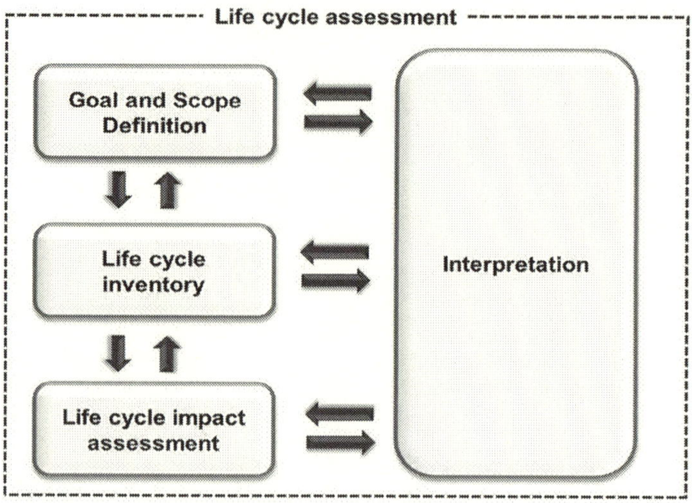

FIG. 1.2 Structure basic for the application of the life cycle assessment methodology.
This figure represents the evaluation structure of the four main stages of the life cycle assessment.

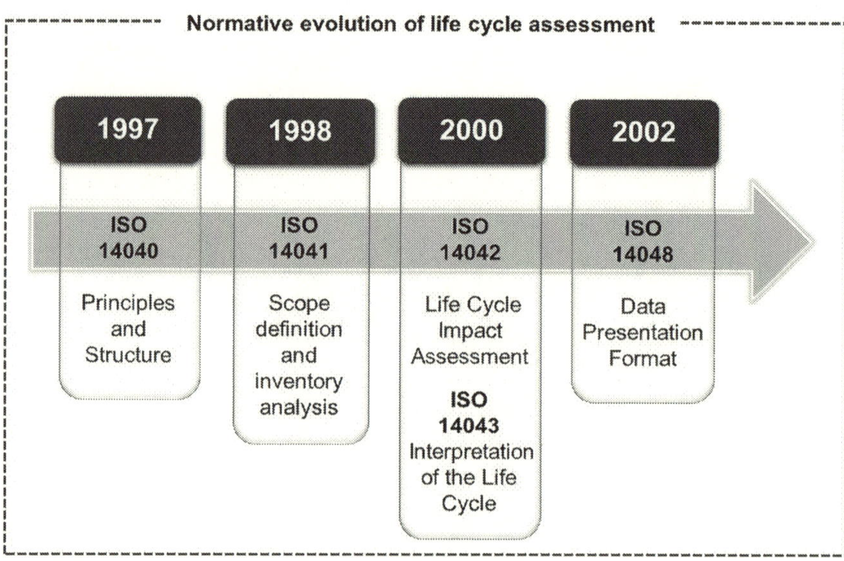

FIG. 1.3 Normative evolution of life cycle assessment.
This figure shows the advancement of the life cycle analysis tool over the years. In addition, it addresses the evolution of the standards for the standardization of the evaluation stages.

References and recommended reading

Papers of **particular interest have been highlighted as:**
• of special interest
•• of outstanding interest

•• Crawford, R. H., Bontinck, P. A., Stephan, A., Wiedmann, T., & Yu, M. (2018). Hybrid life cycle inventory methods—A review. *Journal of Cleaner Production, 172*, 1273–1288. This paper provides an overview of the different hybrid methods of life cycle inventory currently in use. In addition, it presents a detailed critical review, about how each method is applied and its specific strengths and weaknesses.

Guinee, J. B., Heijungs, R., Huppes, G., Zamagni, A., Masoni, P., Buonamici, R., et al. (2010). Life cycle assessment: Past, present, and future. *Environmental Science and Technology*, *45*, 90–96.

Hunt, R. G., Franklin, W. E., & Hunt, R. G. (1996). LCA—How it came about. *The International Journal of Life Cycle Assessment*, *1*(1), 4–7.

Leontief, W. W. (1936). Quantitative input and output relations in the economic systems of the United States. *The Review of Economic Statistics*, *18*, 105–125.

Smith, H. (1963). Cumulative energy requirements for the production of chemical intermediates and products. In *World energy conference*.

Chapter 2

Fundamentals of life cycle assessment: Definitions, terminology, and concepts

The life cycle assessment (LCA) is basically a technique developed with the objective of optimizing methods to better understand and minimize the possible environmental impacts related to the environmental emissions associated with the production and consumption of products from industrial processes and agricultural systems.

More specifically, LCA is a methodology that assesses the environmental burden associated with a product, process, or activity, identifying and quantifying the uses and releases of energy and materials for the environment, and also aims to assess and implement opportunities for improvement. The evaluation includes the entire process life cycle, covering extraction and processing materials; manufacturing, transportation, and distribution; use, reuse, and maintenance; and recycling and final disposal.

This normative, developed by the International Organization for Standardization (ISO), seeks to standardize and describe the structural principles and parameters for conducting and reporting on LCA studies. Thus, the main minimum requirements demanded include (i) the evaluation of environmental aspects and potential process-related impacts by compiling an inventory of relevant inputs and outputs of a system; (ii) the assessment of potential environmental impacts associated with these inputs and outputs; and (iii) the interpretation of the results of inventory and impact assessment phases in relation to the study objectives.

However, for an integral evaluation, it is necessary that the main resources of the required methodology expand, presenting a more complex approach. This way, the main parameters of analysis are (i) to address LCA studies in a systematic and appropriate manner with respect to the environmental aspects of product systems, from raw material acquisition to final disposal, and (ii) the depth of detail and time frame of an LCA study can vary greatly depending on the definition of goal and scope.

Sustainability Metrics and Indicators of Environmental Impact. https://doi.org/10.1016/B978-0-12-823411-2.00010-4

The initial step in choosing the scope and their respective assumptions, methodologies, quality of data collection, and results of LCA studies should be extremely transparent. In addition, LCA studies should discuss and document data sources and communicate succinctly and appropriately. Some steps should be considered, depending on the intended application of the LCA study, to respect confidentiality and property matters (Crawford, 2011). Nevertheless, the LCA methodology should be accessible to the inclusion of new scientific discoveries and improvements in the state-of-the-art technology. Therefore, specific requirements must be applied to the LCA studies that are used to make a statement that is made public (Baumann & Tillman, 2004).

On the other hand, the environmental aspects and potential impacts throughout the life of a product, from the acquisition of raw materials to production, use, and disposal, are considered in LCA studies. In this way, LCA can help in (i) identify opportunities to improve the environmental aspects of products at various points in their life cycle; (ii) decision-making in industry, governmental, or nongovernmental organizations; (iii) selection of relevant environmental performance indicators, including measurement techniques; and (iv) market advertising as a result of an eco-label.

Given this scenario, since LCA adopts a comprehensive and systemic compilation for environmental assessment, ISO validated a standard as part of its 14,000 series, focusing on establishing methodologies for LCA. In this way, a four-stage interactive structure was proposed to perform life cycle analyzes. The four stages include (i) definition and scope of goals, (ii) inventory assessment, (iii) life cycle impact assessment, and (iv) interpretation.

Phases for life cycle assessment

Objective and scope

The first phase of the LCA aims to specify the procedures necessary to define the objective and scope of the study, clearly and plausibly, according to the application in question.

The purpose of the study should represent the reasons for the study and identify the critical points and the application targeted to a specific target audience. Some examples of objectives are the comparison between different products performing the same function and the identification of stages which can be made improvements and new possibilities for the development of new products or processes (EEA, 2000).

In addition, the scope of the study comprises the definition of the system to be studied. It covers, in its entirety, the characteristics of the process performance and the quantification of the functions performed, that is, its functional unit, the establishment of limits of the system, its specifications of data quality requirements, and criteria used for the incorporation or exclusion of input and output information (ISO, 2006).

In this way, it is known that products, processes, or services can take on a variety of functions. Since the results of an LCA must relate only to a functional unit, a comparison of systems is fundamentally reestablished to provide a mathematical standardization of the input and output data. Thus, equivalence between the evaluated systems is promoted.

To elucidate its applicability, the functional unit must be clearly defined and consistent with the purpose and scope of the study. Their inclusion should consist of a quantitative and qualitative description. In addition, its quantification is made through a quantity of product necessary to fulfill a certain function, during a certain period of time. Thus, its definition should include data on product efficiency and durability and performance quality standards.

The limits of the system determine which processes and elementary flows will be included in the LCA and allow to define the level of detail of the study. However, it is well known that studies with excessive detail can result in high economic costs and do not meet the demand for time to complete. Sometimes they may not have great relevance in the results. Moreover, it is noticeable that there are several limits of the system to be specified, being geographical, temporal, technical parameters, and those in relation to natural systems.

As a differential parameter in life cycle assessment, the geographic aspect becomes important at different stages of the life cycle that can occur in different locations. The technologies used and the sensitivity of the environment to the various pollutants may differ from one region to another. Furthermore, it is important to note that these changes occur mainly in the extraction of raw materials.

Given this scenario, the baseline guidelines of the objective and scope were developed in the first stage of life cycle assessment (Fig. 2.1).

First stage: How to set the goal?

At this stage, the purpose and required measures of the study are decided, taking into account the target audience and other key stakeholders. The main objective of this stage is to define the purpose and the application of the study. In a second moment, the study complexity, time, and available resources are parameters that need attention. However, it is relevant to emphasize that the scope of the study and its complexity are strongly influenced by the fact that the life cycle assessment is applied before or after implementation. Thus, when applied to propose a preevaluation, the LCA is prospective, that is, it will present greater assumptions. Therefore, the study will lead to greater knowledge in a period in which it becomes easier to influence the designer and strategies for improvements of the tool.

In short, this step guides the evaluation, so it is based on the approach of how the scenario should be, and the level of competence and detail to be carried out. Thus, it is at this stage that one decides whether the application of LCA will occur in a complete way or only in a small screening.

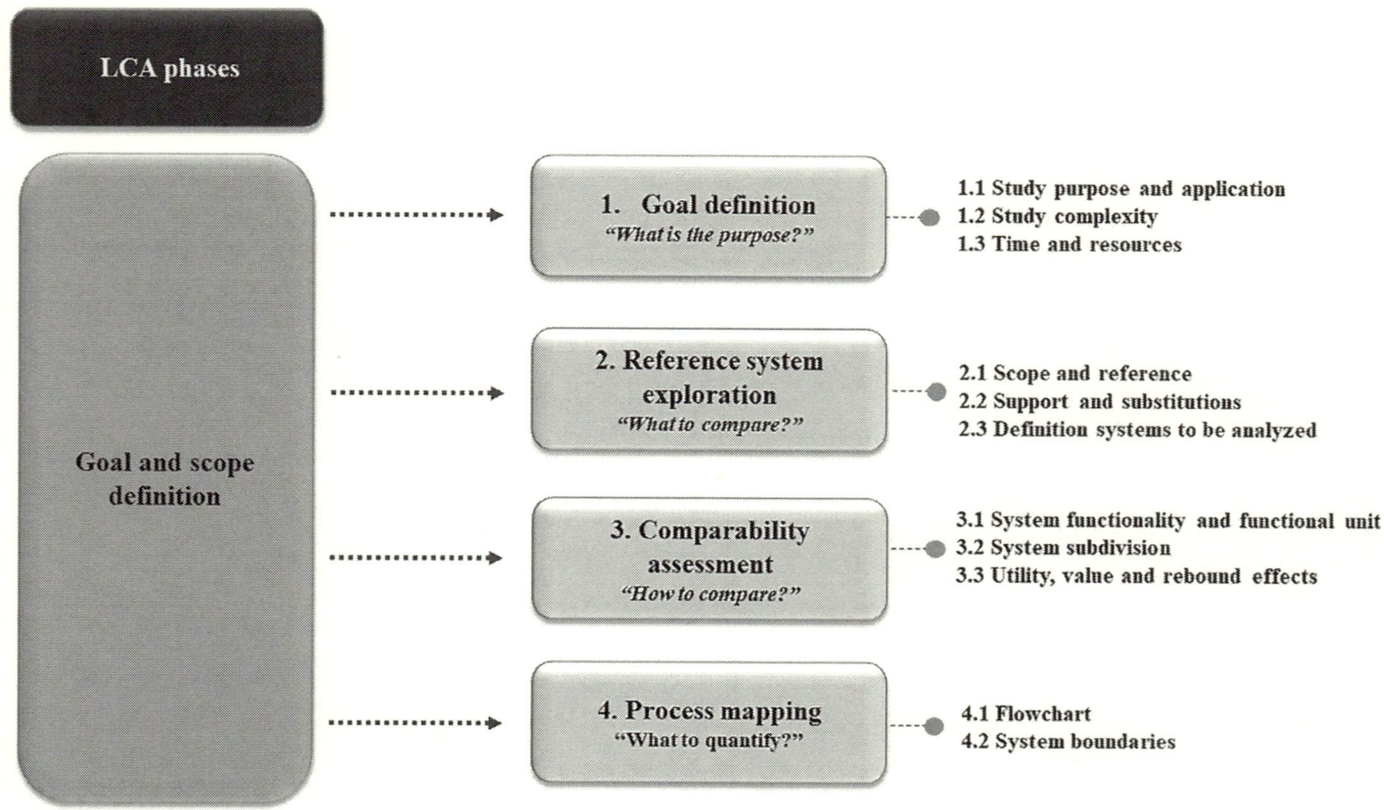

FIG. 2.1 Overview of steps how they link to the phases of an LCA.
This figure reflects a model of questions to be followed to apply the first stage (goal and scope definition) of the life cycle assessment tool.

Second stage: Delimitation of the reference system

The exploration of the reference systems presents the scope of the study. This scope is decided together with the delimitation of the reference system, that is, delineate the system to have with which to compare the study.

This design can occur in relation to three parameters, being the consequences, comparisons, and optimizations of the systems. The consequences are to focus on analyzes that aim to introduce or change a unitary stage of the system, or even the incorporation of a product in the production line of a given process. In contrast, the comparison is intended to compare with a predefined alternative, which may or may not be an exchange or solution option for the system. However, optimization emerges as a parameter to evaluate the different alternatives of possible exchange options.

Third stage: Comparability assessment

This step is designed to ensure that the choices of systems for evaluation are in fact truly comparable and allow for the same functionality. In this step, the functional unit is defined through parameters such as describing the functional result in a measurable way, that is, to indicate quantifications, over a period of time, and if relevant, the location that will be applied. Further, the functional unit should have sufficient scope to capture its role to play within the limits of the systems previously determined. However, it is appropriate not to block the functional unit in a criterion that can be changed as a result. In this way, the number of products and services required to comply with the functional unit must be well defined.

Fourth stage: How to quantify the process mapping

The main objective of this step is to build a general approach to the quantitative data that need to be ascertained. Some examples are visual flowcharts, known in industrial systems as process trees, whose purpose is to expand the view of all processes included, and may expose added processes or processes to be avoided. The best alternative to be implemented are the flowcharts. They have system boundary specifications, delimiting which processes will be submitted for evaluation.

In terms of the reference system, the systems to be evaluated can influence only one stage of the life cycle assessment, resulting in the inclusion within the system boundary. It should be emphasized that the exclusion of significant inputs to the impacts assessed in the LCA can generate inconsistent results in the final analysis.

Nevertheless, although standardization has developed guidelines for the conduct of the first phase of the LCA, many critical points can be identified in the literature. Thus, Table 2.1 presents some application bottlenecks and initial challenges attributed to the starting point of the goal definition and scope.

TABLE 2.1 Guideline requirements for the definition of the objective and scope based on literature review.

Requirement	Main bottlenecks found in the literature	Guideline requirements
#1 Focus on study scoping	The identified challenges are connected to the goal and scope definition phase of an LCA	#1 Guidelines should follow the phases of an LCA, with specific focus on supporting the goal and scope phase
#2 Support reference system definition	Challenge #1: Define the reference system	#2 Guidelines should support an exploration of the reference system to identify relevant substitutions
#3 Support functional equivalence	Challenge #2: Establish functional equivalence between the compared systems in the unit functional definition	#3 The guidelines should support the unit functional definition, which should be accompanied by a comparability assessment, which ensure that: all relevant subfunctions are included; and differences in user perceived value outcome are assessed (to identify rebound effects)
#4 Support system boundary definition	Challenge #3: Ensure a sufficient level of completeness in system boundary selection	#4 The guidelines should support the identification of relevant contributing flows and include "hidden" processes (e.g., from services, infrastructure) in the background system
#5 Support different approaches	Methods range from "full LCA" to simplified/adapted versions of LCA	#5 The guidelines should support both "pre" and "post" assessment and allow for both a "full LCA" approach and more simplified approaches, i.e., a "screening level LCA" based on aggregated data
#6 Focus on scenario approach and qualitative learnings	LCA studies are subject to uncertainty and results usually dependent on assumption about user behavior	#6 The guidelines should support a scenario-based approach and focus should be on the learnings during the evaluation process, not only the quantified results

Based on Kjaer, Pigosso, McAloone, and Birkved (2018).

Life cycle inventory

Life cycle inventory (LCI) is the second phase of the application of life cycle assessment. This step basically involves the data collection and calculation procedures. The information obtained, associated with the inflows and outflows, must be related to the functional unit established in the first phase. It is known that this is the most exhaustive part in performing the life cycle assessment. Still, it is in this phase that depends exclusively on the reliability of the results obtained in the other stages.

Among the components of the life cycle inventory, the product system, the elementary processes, and the data categories stand out (Huijbregts et al., 2001). Product systems are related to one or more defined functions and comprise all the elementary processes. These processes consist of inflows of raw materials and energy, flows of waste output and emissions, and product flows (Fig. 2.2).

Additionally, several procedures are contemplated by ISO 14044 for carrying out the inventory evaluation. Among them, the most important are data collection, data allocation, and validation.

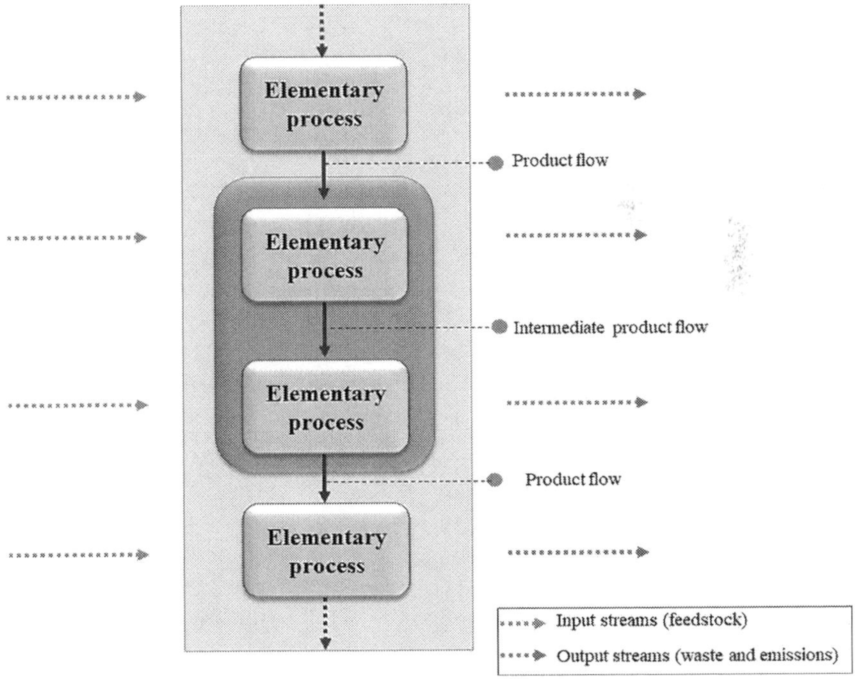

FIG. 2.2 Representation of flow diagram of inventory assessment.
This figure shows an example of a flowchart to be defined, highlighting the inputs and outputs of the system under evaluation corresponding to the life cycle inventory stage.

Preparation for data collection

The purpose is to include the use of simplified flowcharts, which aid in the identification of processes and flows, in the description of each process, and also in the description of the techniques of data collection and calculation for each of the categories of the process.

Data collection

It is known that the data to be collected can be derived from numerous sources, including published scientific literature, the use of software, or direct collection. However, it is important to emphasize that during the data collection process, double counting should be avoided, that is, in the case of multiple flows, it is suggested that allocation methodologies be applied, which should be properly documented.

Allocation

The allocation is one of the most discussed methodological issues in LCA (Finnveden et al., 2009). In this way, there are two main different ways of dealing with multifunctional processes. One is to allocate environmental impacts among products so that whenever possible, subdivision or expansion of the system should be used to avoid allocation problems. If this is not possible, an allocation that reflects the physical (or chemical or biological) causes should be used, and finally, if this is not feasible, allocation based on other measures, such as economic value, mass, or energy, can be used.

Validation of data

It aims to authenticate the real veracity of the data, by establishing comparisons with other mass/energy balances. This step allows verifying if the collected data are, in fact, representative and pertinent for the analysis, the intended use, and purpose.

Conversely, the LCI has evolved significantly in recent decades to become a more systematic and robust tool for sustainability practices. Among some new methods are methods process-based, input/output method, and the hybrid method.

Process-based methods

Process-based modeling aims to compile inventory directly through process analysis. This approach can occur in two ways, being the process flow diagram, which presents the interconnection of flows through a product system. Its quantification is given by simple summation, the extent to which the number of flow inputs is determined by the functional unit and multiplied by the number of environmental practices performed to generate a given product. In addition, the matrix method is known to be more competing, as it presents the entire product system with a wide range of linear equations and solves them simultaneously. In short, this method can be applied to systems with multiple inputs/outputs.

Input/output

The input/output method collects data in the reverse stream, i.e., the complication of the data occurs through the output streams. Thus, since this method analyzes distant upstream stages in the LCI calculation, better-founded results are provided, providing a better system-oriented methodology.

Hybrid method

The hybrid method is a combined methodology of process analysis based on the analysis in input–output. This merged analysis aims to combine the forces of both previous methods. Generally, hybrid methods are free to be widely applied in accordance with the intent of the assessment.

Thus, although many methodologies are proposed and elucidated to improve the reliability of the data collection for the inventory cycle life stage, it is known that some bottlenecks are still appropriate to the application of these methods. In this way, Table 2.2 presents some advantages and guidelines to help in choosing the best method to be used for the intended purpose.

Upstream truncation

It describes the need to delineate a system boundary around the product or process that is under evaluation of parameters regarding time, budget, or data freedom constraints. Yet, can assume that processes of the highest wholeness in the supply chain are generally disregarded. Thus, over the years, as the life cycle inventory databases are discussed in the literature, there is a greater understanding of the supply chain conformations and, therefore, the number of excluded processes tends to be reduced.

Downstream truncation

This level of truncation is classified as the ability of intermediate manufacturing processes to be postponed. That is, they commonly occur between the production of a particular material and its final product. This level is usually attributed in cases of complex products, which require numerous steps in production.

Lateral truncation

Processes are commonly overlooked in a process analysis because of the perceived insignificance of their contribution to the final results, lack of awareness of their existence, or lack of data. This often encompasses nonmaterial processes, such as the provision of services, or physical inputs considered small enough to be excluded.

TABLE 2.2 Advantages and bottlenecks of choosing LCI methods.

LCI	Methodology	Advantages	Bottlenecks	Suitable application
Process-based methods	Process flow diagram	Necessary level of detail Best result for single product system Easier to understand	Not suitable for larger system/multiple input/output/recycling Time consuming for huge data Truncation error	LCI for single product system (Kulak, Nemecek, Frossard, Chable, & Gaillard, 2015) LCI for production chain without recycling (Suh & Huppes, 2005) Sustainable process selection (Nucci, Puccini, Pelagagge, Vitolo, & Nicolella, 2014)
	Matrix method	Effective for larger system/multiple input/output/recycling Necessary level of detail	Time consuming for huge data Truncation error Mathematical expertise required	Raw material selection (Ocampo et al., 2015) Process selection/redesign (Wang & Work, 2014)
Input/output methods	Input/output Life cycle assessment	No requirements of unit process data Calculate environmental impacts The probability of lower error	Does not cover the entire life cycle Lack of detail Uncertainty of data Data update Unappropriated for import	Decision-making (De Benedetto & Klemeš, 2010) LCI for larger system boundary (Tan, Aviso, Barilea., Culaba, & Cruz, 2012)
Hybrid methods	Hybrid	Basic More complete boundary than process-based modeling Lower data uncertainty	Double score The interaction between process-based and input/output based is not validate Loss of dynamic representations	Life cycle inventory in case of imports/exports LCI until product consumption stage (Junnila & Horvath, 2003)

On the other hand, considering the analysis of the inventory to the stage with the major risk of suitability in the data, a variety of software was developed to assist this tool, being used as facilitators, besides having databases of elements common to several cycles of life (Table 2.3).
Based on Islam, Ponnambalam, and Lam (2016).

TABLE 2.3 LCA software tools and LCI databases around the world.

LCA software tools		LCI database			
Name	*Focus*	*Name*	*Description*		
			Source	*Region*	*Focus*
SimaPro	Generic LCA tool; evaluate environmental performance of products and services; its Eco-Invent contains 2500+ process database	Eco-Invent	Primary, BUWAL, ETH-ESU	Swiss and Western Europe	Generic; over 2500+ processes; includes uncertainty data and infrastructure
ATHENA	Contain North American database; specifically developed for buildings; allow user for comparisons among alternative	Athena's database	Primary	Canada and North America	Construction industry; 90+ processes—wood, steel, concrete, and structural products
GaBi	Generic LCA tool; process-based LCA; widely adopted tool integrated with economic cost	USA National LCI	Primary, Eco-Invent 2000	USA and North America	Basic processes to build upon in LCA Studies
LISA	Australian-based freely available streamlined tool; not allow users to add new case studies	Australian LCI	Primary, various databases	Australia	Mostly AusLCI and LCA studies at RMIT University/center for design and others
LEGOE	Use by architects and engineers; integrated with a CAD tool	Canadian database	Primary	Canada and North America	Basic materials: aluminum, glass, plastics, steel, and wood
BEES	US-based building material specification tool; applicable to LCA and LCC; not transparent	Inventory of Carbon & Energy (ICE)	Secondary	Global	Inventory of embodied energy and carbon coefficient for building materials

Additionally, in a critical view, although the most stable way to collect data is through the use of software, it is understood that, regardless of the methodology of data selection, process analysis usually has the prerogative to use specific data to the product under study in order to allow the highest level of feasible accuracy. However, such systems are susceptible to systemic incompleteness, as it is unfeasible to extensively assess the supply chain of processes due primarily to cost and time constraints. Given this scenario, essential levels of truncation were defined for process analysis, being truncation upstream, downstream, and lateral.

Based on Islam, Jollands, and Setunge (2015).

Life cycle impact assessment

Based on the ISO standards, the life cycle impact assessment is the third step of the LCA tool, which represents the quantitative or qualitative process applied in the characterization and evaluation of the impacts associated with the life cycle inventory. Throughout this stage, flows of inputs and outputs of materials and energy are converted into results of environmental impacts (De Haes et al., 2002).

During the implementation of the life cycle impact assessment (LCIA), there are three mandatory steps, which are the selection of impact categories, the classification of product categories into different impact categories, and the characterization of possible impacts, estimating and using some scientific factors (Sonnemann, Castells, Schuhmacher, & Hauschild, 2004). On the other hand, optional elements such as standardization, aggregation, and weighting can be applied to provide additional information, which guarantees an LCA profile with greater precision of analysis (Fig. 2.3).

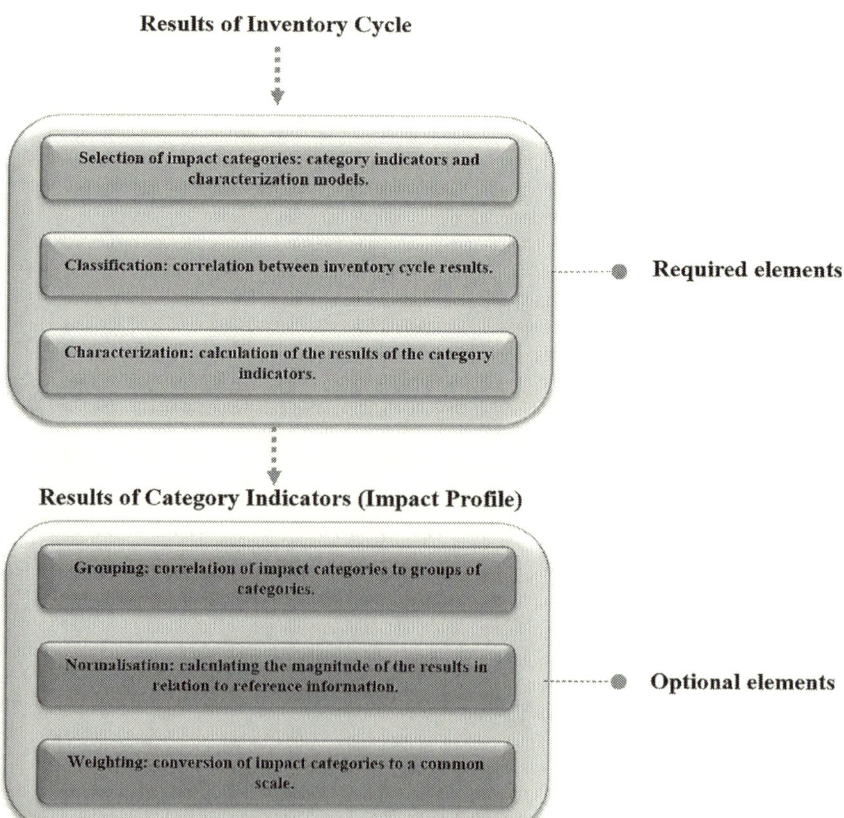

FIG. 2.3 Required and optional elements of life cycle impact assessment.
This figure represents the principal and optional elements of the life cycle assessment tool.

Among the elements required, the selection of impact categories consists of the first choice and decision-making, when defining and delimiting their indicators (Pennington et al., 2004). The principle of this selection must be based primarily on scientific knowledge of the environmental mechanisms related to the analyzed process. For some examples of the categories of impact used in existing methods, we can mention climate change, reduction of stratospheric ozone, acidification, use of natural resources, and land use (Hauschild & Huijbregts, 2015).

The second required element is known as the classification step. This step is defined as the correlation between the results obtained in the inventory analysis against the different impact categories established in the previous step. Still, it is important to emphasize that the classification stage is a qualitative process, totally based on the scientific analysis of environmental aspects.

Additionally, in the characterization stage, the results of the indicators for each impact category are quantified through mathematical modeling (Eq. 2.1). Unlike classification, this is a quantitative process and requires great attention from the analysts involved.

$$C_i = \sum CF \times EI \tag{2.1}$$

where C_i is the impact category, CF is the characterization factor, and EI is emission inventory, expressed in mass released into the environment per functional unit.

Furthermore, the characterization process can adopt behaviors such as the marginal, linear, and average approaches. In short, the marginal approach consists of being the standard behavior to derive the characterization factors based on accounting for minimum changes in the elementary flows. However, the linear approach is defined by the simplified behavior, since the analysis assumes surface character. Finally, the medium approach aims to analyze long-term behavior.

On the other hand, to obtain a life cycle assessment with the greater scope of comparison and less data inconsistency, optional parameters such as normalization, grouping, and weighting are encouraged to be used.

Normalization is defined for the benefits of placing the indicated impact indicator in a broader context. In principle, the use of the normalization scores in LCA case studies is relatively straightforward. It is expressed in a way that allows the impact indicators to be compared to each other, such that, the sum of each category indicator result is divided by a reference value according to Eq. (2.2) as follows:

$$N_K = \frac{S_K}{R_K} \tag{2.2}$$

where k denotes the impact category, N is the normalized indicator, S is the category indicator from the characterization phase, and R is the reference value or the normalization factor, expressed in mass of emission per capita per year.

Additionally, normalization factors are usually chosen to represent the actual or potential magnitude of the corresponding impact category for a geographic area and over a certain period of time, i.e., the standardization factors involved represent the potential impact of that category. At the exemplary level, the total annual emissions of a reference year in a region are normally used to calculate the normalization through published environmental reports for the global warming climate, or other attributes as reference values related to total emissions or use of natural resources for a preestablished baseline.

Weighting process consists of multiplying the weighting factors by the standardization result (values normalized) for each impact category. The potentials for different impacts are evaluated according to their severity and, according to the result, different weighting factors are established: one factor for each category of impact and damage. The weighting is often applied in the form of linear weighting factors according to Eq. (2.3):

$$Wg_k = \sum WgF_k \times N_k \tag{2.3}$$

where Wg is the impact weighting, WgF_k is the weighting factor for impact category k, N_k is the normalized indicator, and k is the impact category indicator, generally expressed in percentage.

Currently, since weighting is an optional attribute, its methodology is not uniformly standardized for weighting individual indicators before aggregating them into a single composite indicator. Among the different methods that can be attributed, they can be classified in similar, statistical models or based on public opinion (Table 2.4). However, equal weights criteria are often applied because they have a significant value in the indicator impact and in their respective classification, especially when a greater weight is given to indicators in which some countries surpass or fail. Thus, regardless of the procedure used, all weighting models need to be transparently explained.

TABLE 2.4 Different weighting methods and their principles of application.

Classification	Criteria	Principle of the method
Similar	Equal weights	The variables receive the same weight when there is no statistical or empirical basis for choosing a different scheme. Equal weighting may imply recognition of equal status for all indicators (for example, when policy evaluations are involved)
Statistical models	Principal component analysis	They group indicators that are collinear to form a composite indicator capable of capturing as much information as possible from these indicators. Each factor reveals the set of indicators with the largest association with it

TABLE 2.4 Different weighting methods and their principles of application—cont'd

Classification	Criteria	Principle of the method
	Data envelopment analysis	Uses linear programming to retrieve an efficiency frontier and uses it as a benchmark to measure the performance of a given set of countries. Positive weights should be considered (the higher the indicator value, the better for the corresponding country); nondiscrimination of countries that are better in any dimension (i.e., indicator), classifying them equally; linear combination of the best performances is feasible (border convexity)
	Regression analysis	Correlations between a large number of indicators and a single output measure that represents the goal to be achieved. This approach, although adequate for a large number of indicators of different types, implies the assumption of linear behavior and requires the independence of explanatory indicators
	Component models not observed	They are obtained through the likelihood of the indicators. This method resembles the linear regression approach, but its main difference is that the dependent variable is unknown
Public opinion	Budget allocation	The specialists receive a budget of numerous points, which are placed in several indicators, paying more for what they themselves want to emphasize
	Public opinion/ specialist	Participatory approaches, involving public or expert judgment, are often used to determine weights, in order to express the relative importance of indicators from the societal point of view
	Public opinion	Opinion polls focus on the notion of concern, that is, people are asked to express "too much" or "little concern" about certain problems measured by indicators
	Analytic hierarchy process	Used in decision-making of several attributes. It allows the decomposition of a problem into a hierarchy and ensures that qualitative and quantitative aspects of a problem are incorporated into the evaluation process, during which opinion is systematically extracted through peer comparisons, raising the question "which of the two indicators is more important? 'Secondly' by how much?"

Continued

TABLE 2.4 Different weighting methods and their principles of application—cont'd

Classification	Criteria	Principle of the method
	Conjoint analysis	Joint analysis is a multivariate and decomponential data analysis technique, commonly used in marketing and consumer research. This methodology requires an analysis (priority) in a grouping of alternative scenarios. This priority is then decomposed by associating the individual components (the known values of the indicators in that scenario) to the evaluation

Also, in the impact evaluation stage, there is optionally the grouping parameter. This parameter consists of the conjuncture of similar impact categories or the classification of impact categories in protection topics. The objective of this step is to help through the grouping of the evaluation of the environmental performance.
Based on Ludlow (2018).

Interpretation

The fourth and final stage of life cycle assessment is denominated interpretation. This phase is known to be one of the most sensitive stages, as the assumptions made during the earlier stages of inventory and impact assessment, as well as the adjustments that may have occurred, may affect the final outcome of the project. In addition, this stage is responsible for formulating recommendations and conclusions to be presented to the target public through the final reports.

The preparation of the final report should be described in such a way as to make it possible to use the results and their interpretation according to the objectives established for the study. It should consist of the steps of identification of significant issues, evaluation and verification, and conclusions, recommendations.

Identification of significant issues

The identification and structuring of information, such as methodological choices, allocation criteria, models and characterization factors, and category indicators, are performed. In addition, some mechanisms may include grouping and structuring the outcomes of the LCI and LCIA steps together with data quality, identifying the value choices used in the study, and the roles and responsibilities of different stakeholders in the application. In this way, the significance of data categories, as well as the categories of impact and processes included in the processes, is established.

TABLE 2.5 Verification criteria for the evaluation of LCA final reports.

Verification models	Finality
Integrity	Ascertain the lack of data in the inventory or completeness of the impact assessment
Consistency	Evaluates whether modeling and methodological choices are appropriate and consistent with the purpose and scope of the study
Uncertainty	Revises the effect of the uncertainty of the data on the result obtained
Sensitivity	Identifies the effect of critical data
Quality of data	Evaluates the degree of lack of data, approximate information, and appropriate data

Based on ILCD (2010).

Evaluation

To assess the reliability of the study results, the relevance, availability, and integrity of the data should be ensured. However, some assumptions of uncertainties both in methods and in data, in relation to the objective and scope, can be verified in this criterion. Thus, Table 2.5 shows the purposes of the applied verification models.

Outcomes conclusion

At this stage, the conclusions are outlined, which should be consistent with the purpose and scope of the study and with the data quality requirements and assumptions made. The recommendations made must be in accordance with the final conclusions of the study and with the intended application.

However, the complexity of the interpretation depends directly on whether the purpose of the evaluation is to direct improvements, to recommend strategies for prioritizing project choice efforts, or to determine a more in-depth life cycle assessment.

Boundaries for life cycle assessment

Given the wide range of application life cycle assessment for industrial processes and products, different scenarios are approached to adapt the mathematical tool for major precision in the obtention of results. In this context, some alternatives to life cycle assessment were proposed to better characterize the analysis adopted. However, it should be emphasized that the changes in the life cycle assessment, when compared with the alternative life cycle assessment, are exclusively focused on the delimitation of the system to be studied. Thus, to date, five alternatives have been defined: cradle to grave, cradle to factory gate, cradle to cradle or production of open mesh, the gate to the gate, and the wheel well (Fig. 2.4).

Life cycle assessment from cradle to grave

Evaluation of the life cycle from the cradle to the grave is defined as the most complete analysis of the LCA. This delimitation of the systems involves the extraction of raw materials, through its phase of use and ending in its final disposal or phase of elimination (grave).

Life cycle assessment of cradle to gate (or farm)

The life cycle assessment from the cradle to gate consists of the application of the tool from the extraction of natural resources, the crib, to the gate of the factory or farm, that is, the delimitation of the system is defined until before it is transported to the consumer. Thus, the phases of use and final disposal of the products or machinery involved in the production system are not involved in the quantifications of the system, resulting in the omission of these data.

Generally, these cradles to factory gate evaluations are sometimes applied in base reports derived from environmental product claims, termed business-to-business product environmental statements.

Assessment of cycle from cradle to cradle or production of open mesh

The life cycle assessment of the cradle to the cradle, or also called open production, is defined as a specific model of the life cycle analysis from cradle to grave. However, its differential is attributed in the final disposal stage, where the recycling process is considered.

This method is used to reduce the environmental impact of processes, applied to production, operation, and sustainable disposal practices. Moreover, it is possible to aim for the incorporation of social responsibility during the development of processes. In this way, the recycling processes originate new identical products or different products. However, the allocation of product loads to open-loop production systems presents considerable challenges to LCA, a commonly applied example are methods such as the avoided burden approach which is intended to address these issues.

Life cycle assessment from the gate to the gate

The life cycle assessment from the gate to the gate has a partial view, presenting only a value added in the process of the entire production chain. This modality can then be linked in its proper production chain to form an evaluation of the crib to the full gate.

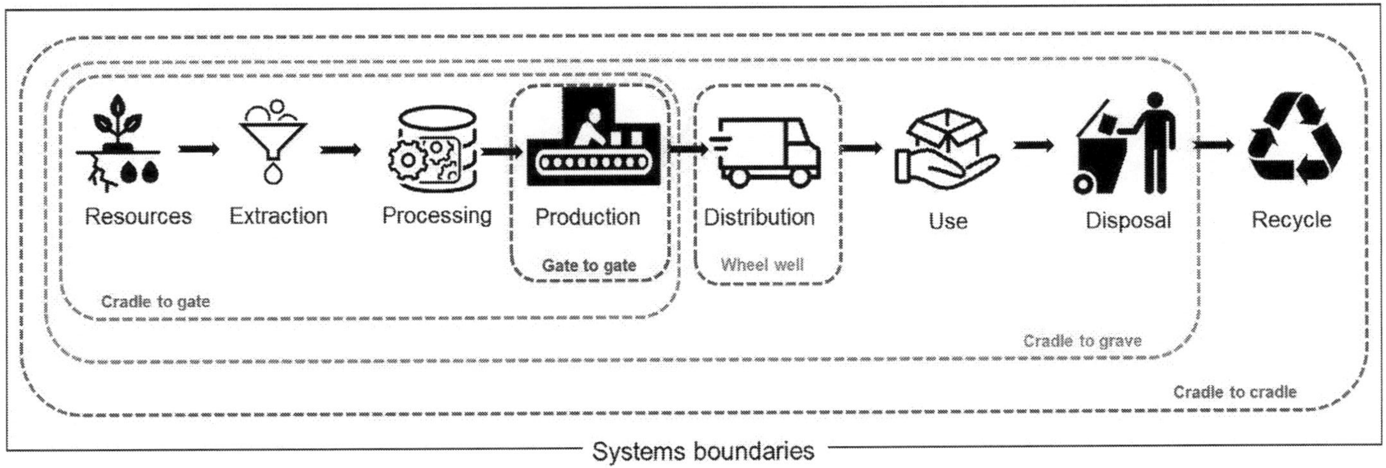

FIG. 2.4 Framework general for systems boundaries delimitation.
This figure shows, through a generic flowchart, some examples of definitions of the system boundaries.

Life cycle assessment of the wheel well

Life cycle assessment from well to wheel is applied exclusively for the transportation of fuels and vehicles and engines. Divided into two stages, the first stage is responsible for incorporating the production of raw material, processing, and delivery of fuel often referred to as upstream. In contrast, the second stage is the operating state of the vehicle, which itself is sometimes referred to as downstream.

In addition, this analysis is generally applied to assess the total energy consumption, energy conversion efficiency, and, above all, the emissions impacts generated by the transportation of products.

On the other hand, by including the carbon footprint, this variant of life cycle assessment aggregate models is developed by national and international laboratories to quantify the impacts of new fuels and vehicle technologies. These energy use reports typically report on greenhouse gas emissions as the volatile organic compounds (VOCs), carbon monoxide (CO), nitrogen oxides (NO_x), and sulfur oxides (SO_x).

References and recommended reading

Papers of particular interest have been highlighted as:
• of special interest
•• of outstanding interest

• Baumann, H., & Tillman, A. M. (2004). *The Hitch Hiker's guide to LCA: An orientation in life cycle assessment methodology and applications*. Student Literature. This book presents a substantial approach of methodological alternatives to life cycle assessment, their implications, and their usefulness in various applications such as product development, marketing, production, and waste management.

• Crawford, R. (2011). *Life cycle assessment in the built environment*. Routledge. This book presents case studies using the new hybrid approach to quantify the inflows and outflows of life cycle inventories. Further, the studies in this book enable a comprehensive analysis of a wide range of environmental resource and product requirements so that the possible environmental impacts of a building system or infrastructure can be verified.

•• De Haes, H. U., Finnveden, G., Goedkoop, M., Hertwich, E., Hofstetter, P., Klopffer, W., et al. (2002). Life cycle impact assessment: Striving towards best practice. In *SETAC Press proceedings*. This publication focuses on furthering the development of a technical framework, offers an overview of existing data and methods for different impact categories, and explores evaluation criteria, including scientific validity, transparency, environmental relevance, feasibility, and links with life cycle inventory.

• EEA. European Environment Agency. (2000). *Environment in the European Union at the turn of the century: Offprint: Greenhouse gases and climate change*. Office for Official Publ. of the European Communities. This document reports the guidelines are supported by a number of technical reports, and which present the scientific discussions and documentation for recommendations in the field of the life cycle assessment.

• Finnveden, G., Hauschild, M. Z., Ekvall, T., Guinée, J., Heijungs, R., Hellweg, S., et al. (2009). Recent developments in life cycle assessment. *Journal of Environmental Management*, *91*(1), 1–21. This paper presents a review of recent developments in LCA methods. In addition, the focus of the study focuses on the distinction between attributional and consequential life cycle assessment. Furthermore, a critical analysis is carried out in relation to the strengths and weaknesses of the LCA, as well as questions regarding the categories of impact and weighting.

Hauschild, M. Z., & Huijbregts, M. A. (2015). Introducing life cycle impact assessment. In *Life cycle impact assessment* (pp. 1–16). Dordrecht: Springer.

Huijbregts, M. A., Norris, G., Bretz, R., Ciroth, A., Maurice, B., von Bahr, B., et al. (2001). Framework for modelling data uncertainty in life cycle inventories. *The International Journal of Life Cycle Assessment*, *6*(3), 127.

● ILCD. (2010). *ILCD handbook: General guide for life cycle assessment—Detailed guidance* (1st ed.). Luxembourg: Institute for Environment and Sustainability, Joint Research Centre, European Commission. This document provides technical guidance for detailed life cycle assessment studies and provides the technical basis to derive product-specific criteria, guides, and simplified tools.

● International Organization for Standardization (ISO). (2006). *Environmental management: Life cycle assessment; Principles and framework*. (No. 2006). Life cycle assessment standard, this document describes the principles and structure of life cycle assessment, including all phases of assessment, as well as reports and critical review of stroke, and your limitations. In addition, it also reports the relationship between the phases of LCA and conditions of use of value choices and optional elements.

Islam, H., Jollands, M., & Setunge, S. (2015). Life cycle assessment and life cycle cost implication of residential buildings—A review. *Renewable and Sustainable Energy Reviews*, *42*, 129–140.

Islam, S., Ponnambalam, S. G., & Lam, H. L. (2016). Review on life cycle inventory: Methods, examples and applications. *Journal of Cleaner Production*, *136*, 266–278.

Kjaer, L. L., Pigosso, D. C., McAloone, T. C., & Birkved, M. (2018). Guidelines for evaluating the environmental performance of product/service-systems through life cycle assessment. *Journal of Cleaner Production*, *190*, 666–678.

Ludlow, P. (2018). The European Commission. In *The new European community* (pp. 93–140). Routledge.

●● Pennington, D. W., Potting, J., Finnveden, G., Lindeijer, E., Jolliet, O., Rydberg, T., et al. (2004). Life cycle assessment Part 2: Current impact assessment practice. *Environment International*, *30*(5), 721–739. This paper highlights how practitioners and researchers from many domains have come together to provide indicators for the different impacts attributable to products in the life cycle impact assessment phase.

Sonnemann, G., Castells, F., Schuhmacher, M., & Hauschild, M. (2004). Integrated life-cycle and risk assessment for industrial processes. *The International Journal of Life Cycle Assessment*, *9*(3), 206–207.

Kulak, M., Nemecek, T., Frossard, E., Chable, V., & Gaillard, G. (2015). Life cycle assessment of bread from several alternative food networks in Europe. *Journal of Cleaner Production*, *90*, 104–113.

Suh, S., & Huppes, G. (2005). Methods for life cycle inventory of a product. *Journal of Cleaner Production*, *13*(7), 687–697.

Nucci, B., Puccini, M., Pelagagge, L., Vitolo, S., & Nicolella, C. (2014). Improving the environmental performance of vegetable oil processing through LCA. *Journal of Cleaner Production*, *64*, 310–322.

Ocampo, L., Carreon, R., Carvajal, J. A., Galagar, K. J., Gialolo, D. M., Gulayan, M., et al. (2015). Matrix-based life cycle assessment (MLCA) on polystyrene and recycled paper egg tray packaging. *Journal of Production Engineering*, *18*. 87–81.

Wang, R., & Work, D. (2014). Application of robust optimization in matrix-based LCI for decision making under uncertainty. *The International Journal of Life Cycle Assessment*, *19*(5), 1110–1118.

De Benedetto, L., & Klemeš, J. (2010). The environmental bill of material and technology routing: An integrated LCA approach. *Clean Technologies and Environmental Policy*, *12*(2), 191–196.

Tan, R. R., Aviso, K. B., Barilea., I. U., Culaba, A. B., & Cruz, J. B., Jr. (2012). A fuzzy multiregional input–output optimization model for biomass production and trade under resource and footprint constraints. *Applied Energy*, *90*(1), 154–160.

Junnila, S., & Horvath, A. (2003). Life-cycle environmental effects of an office building. *Journal of Infrastructure Systems*, *9*(4), 157–166.

Chapter 3

Methods of evaluation of the environmental impact on the life cycle

To assess the potential impacts generated by elemental flow quantifications in the inventory analysis, impact categories arise as a class of environmental issues used to characterize, evaluate, and compare the systems under evaluation.

It is known that in the life cycle impact assessment phase, different methods are used to calculate these impacts, however, the main difference between the methods is concentrated in the mid- and endpoints (Hauschild et al., 2013). In short, the midpoints are responsible for analyzing the environmental impacts located along the chain of cause and effect, that is, before the endpoint is reached. On the other hand, endpoint methods examine the environmental impact at the end of this chain. For example, the midpoint will assess the increased concentration of greenhouse gases in the atmosphere, while the endpoints will assess its potential impacts on climate change.

Thus, according to Fig. 3.1, the impact categories of average points include ozone depletion potential, photochemical oxidation potential, ecotoxicity, acidification potential, eutrophication potential, net energy ratio, water footprint, land use, global warming potential, and human toxicity, while the most relevant endpoints with the receiving category are ecosystem quality, climate change, and human health (Jolliet et al., 2003).

However, while ISO standards present an extensive list of impact categories, the selection of these categories is directly related to the purpose of the study. More precisely, it is clear that the selection criteria will refer to the type of application as well as the conclusions in the decision-making (Tillman, 2000). Furthermore, there is a consensus among international applicators that the classification of impact categories, such as geographic scale type or characterization methodology, is a responsibility of the practitioner, and currently, no methodology integrates specific guidelines as to which categories should be life cycle assessment (Table 3.1).

Sustainability Metrics and Indicators of Environmental Impact. https://doi.org/10.1016/B978-0-12-823411-2.00003-7

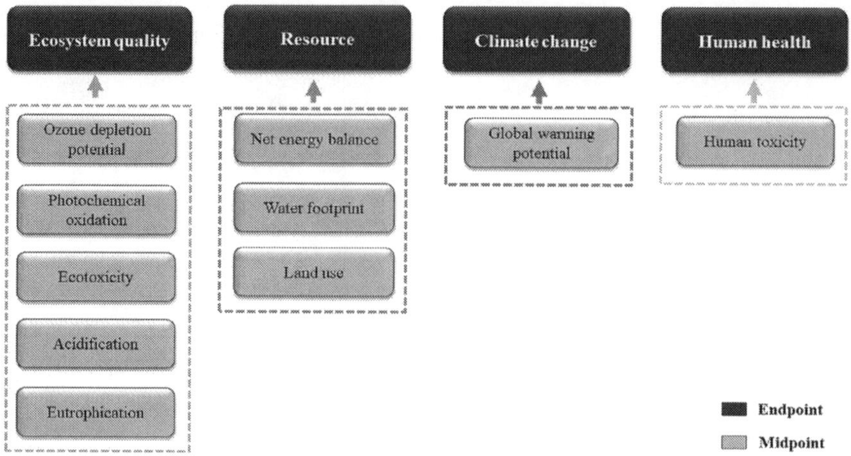

FIG. 3.1 Impact categories according to the midpoints and endpoints model. This figure illustrates the main impact categories and their respective points.

TABLE 3.1 Main categories of impact and the consensus at the international level.

Impact categories	Geographic scale			International consensus	
	Global	*Regional*	*Local*	*Classification*	*Characterization*
Ozone depletion potential	•			Yes	Yes
Photochemical oxidation potential		•		Yes	Yes
Ecotoxicity		•	•	Yes	No
Acidification potential		•	•	Yes	Yes
Eutrophication potential		•	•	Yes	Yes
Global warming potential	•			Yes	Yes
Land use			•	Yes	No
Human toxicity		•	•	Yes	No
Resource consumption (net energy ratio, water footprint)	•	•	•	Yes	Yes

Impact categories

Ozone depletion potential (stratospheric)

Stratospheric ozone acts as an important filter to protect the earth against ultraviolet radiation. This category represents potential damage to the stratospheric ozone layer caused by entry enhanced by both anthropogenic compounds and halogenated compounds.

To select which substances actually affect the ozone layer, aspects such as being sufficiently stable in the atmosphere to allow a substantial fraction to reach the stratosphere, as well as substances containing chlorine or bromine, which upon release into the atmosphere, participate in a chemical decomposition of ozone, were established as the main contributing emissions (Tanaka, Oldfield, Neece, Mullins, & Allen, 2000).

Given this scenario, halogenated compounds, for example, chlorofluorocarbons (CFCs) and their derivates (CFC-11, CFC-12, CFC-113, and CFC-114), hydrofluorocarbons (HFCs), halons, etc., because they have relatively stable characteristics and because they are decomposed into chlorinated and brominated substances under the influence of ultraviolet radiation (UV), represent as potentially harmful pollutants to the atmosphere. In addition, other substances such as nitrogen oxides, hydrogen oxides, methane act as catalysts in the decomposition of ozone. Therefore, the specific characterization of the substances varies for each characterization factor (Table 3.2).

TABLE 3.2 CFC-11 equivalence factors of substances relative to ozone depletion.

Compounds	kg CFC-11 equivalence factor
1 kg CFC-11	1.0 kgeq CFC-11
1 kg CFC-12	0.82 kgeq CFC-11
1 kg CFC-113	0.90 kgeq CFC-11
1 kg CFC-114	0.85 kgeq CFC-11
1 kg CFC-115	0.40 kgeq CFC-11
1 kg Halon 1301	12 kgeq CFC-11
1 kg Halon 1211	5.1 kgeq CFC-11
1 kg Halon 2402	6.0 kgeq CFC-11
1 kg Carbon tetrachloride (CCl_4)	1.20 kgeq CFC-11
1 kg Methyl chloroform (CH_3CCl_3)	0.12 kgeq CFC-11
1 kg HCFC-22	0.034 kgeq CFC-11

Based on EEA (2000).

The major consequence of ozone decomposition is the increase in UV radiation received (especially UV-B radiation), leading to impacts on humans such as increased levels of skin cancer potential, cataracts and decreased immune defense. Yet, the increase in UV radiation also has an impact on organisms and ecosystems mainly where the decomposition of the ozone layer is already significant (Hauschild & Wenzel, 1998).

Quantitatively, as an impact of a given process, ozone depletion potential (stratospheric) can be estimated by summarizing ODPs according to Eq. (3.1):

$$ODP = \sum ODP_i \times m_i \tag{3.1}$$

where ODP is the ozone depletion potential, ODP_i is the equivalency factor for the substance i (CFC-11-eq), and m_i is the mass of the emission substance i (kg), expressed in kg CFC-11-eq.

Therefore, a brief ozone depletion potential category outline can be seen in Table 3.3, and a calculation exemplification is inserted in Box 3.1.

TABLE 3.3 Ozone depletion potential (stratospheric) category outline.

Impact category	Depletion of ozone (stratospheric)
Definition	Diminution of the stratospheric ozone layer due to anthropogenic emissions of ozone depleting substances
Impact indicator	Increase in UV-B radiation and the number of cases of skin diseases
Considerations	Atmospheric residence time of ozone depleting substances
Damage categories	Ecosystem quality and human health
Unit	kg CFC-11 equivalent

BOX 3.1 Calculation application example for ozone depletion potential category.

As can be seen through Eq. (3.1), the calculation of the evaluation of the ozone depletion potential can be quantified by multiplying the mass quantity emitted by the process, related to the equivalence factor of trichlorofluoromethane.

For example, the use of fossil electric energy based on coal has an equivalency factor of approximately 9.39×10^{-8} kg CFC-11 eq for each 1 kWh. Note that equivalence factors for electric energy are generally found in IEA reports. Thus, supposing that a textile industry system uses approximately 1.38 kWh for each 1 kg of textile, and the average monthly production is 85,550 kg of textile, therefore:

BOX 3.1 Calculation application example for ozone depletion potential category—cont'd

$$ODP = \sum OPD_i \times m_i$$

$$ODP = 9.39 \times 10^{-8}\,\text{kg CFC} - 11_{eq} \times 85{,}550.0\,\text{kg}$$

$$ODP = 8.03 \times 10^{-3}\,\text{kg CFC} - 11_{eq}$$

Therefore, the ozone depletion potential for this process is estimated in 8.03×10^{-3} kg CFC-11 eq. Note that the importance of quantifying this category is in the ability to assess the ozone depletion potential caused by GHG emissions, leading to a greater proportion of solar UV-B radiation reaching the Earth's surface.

Photochemical ozone creation potential (tropospheric)

The photochemical ozone creation potential (POCP) can be understood by the degradation of ozone formed in the troposphere under the influence of sunlight when nitrogen oxides are present. However, in the presence of the volatile organic compounds, the peroxide radicals can also be produced. Still, peroxide radicals are highly reactive and toxic compounds, and the presence of peroxide radicals can result in an increase in ozone concentration through a complex reaction standard.

On the other hand, since ozone is a secondary pollutant, and since there are practically no ozone sources of emissions from human activity, it has been clarified that the main precursors of this pollutant are nitrogen oxides (NO_x), volatile organic compounds (VOCs), including methane (CH_4) and carbon monoxide (CO).

The crucial consequences of this photochemical generation of ozone may be directly related to the biochemical effects of reactive ozone compounds when exposed to vegetation, the extent to which photosynthetic damage can occur and cause total plant death. However, exposure of humans to ozone can result in irritation, respiratory problems, and chronic damage to the respiratory system (Fenger, 2009).

In this context, the POCP is defined as the additional ozone formed in a modeling resulting from a time interval when adding a certain amount of volatile organic compound in relation to the addition of the same mass of ethylene expressed in mass of ethene equivalents by mass of gas (kg C_2H_4/kg VOC) (Eq. 3.2). It can be classified as smog or tropospheric ozone. In sum, the former is characteristic of local impacts when quantified in low proportions, while the latter is defined as encompassing a regional scale.

$$POCP = \frac{OI_{VOC}}{OI_{ethene}} \tag{3.2}$$

where *POCP* is the photochemical ozone creation potential, OI_{VOC} is the ozone increment with VOC (kg), and OI_{ethene} is the ozone increment with ethene (kg), expressed in kg C_2H_4/kg VOC.

Therefore, a brief photochemical ozone creation potential category outline can be seen in Table 3.4, and a calculation exemplification is inserted in Box 3.2.

TABLE 3.4 Photochemical ozone creation potential category outline.

Impact category	Photochemical ozone creation
Definition	Type of smog created from the effect of sunlight, heat, and nonmethane volatile organic compounds (NMVOCs) and NO_x
Impact indicator	Increase in the summer smog
Considerations	Based on meteorology information, beyond the chemical composition of the atmosphere and emissions of other pollutants
Damage categories	Ecosystem quality and human health
Unit	kg ethylene (C_2H_4) equivalent

BOX 3.2 Calculation application example for photochemical ozone creation potential category.

The generation of volatile organic compounds (VOCs) is one of the main sources of pollution in composting facilities, from which undesirable odors are released. In addition, the generation of photochemical smog with other gases, such as NO_x, can produce ozone at ground level due to its potential for photochemical ozone creation (POCP).

VOCs are also responsible for the generation of photochemical smog due to its potential for creating photochemical ozone (POCP). Specifically, POCP results from the reaction of VOCs in the presence of NO_x under the influence of solar radiation, which produces tropospheric ozone (Guinée & Lindeijer, 2002). The efficient photodissociation of NO_2 is catalyzed by hydroxyl radicals (radical point OH) and generates ozone according to the following reactions (Jenkin & Hayman, 1999):

$$NO_2 + \text{sunlight}\,(< 420\,\text{nm}) \rightarrow NO + O \tag{R1}$$

$$O + O_2 \rightarrow O_3 \tag{R2}$$

The hydroxyl radical is the last stage of the sequence $RO_2 \rightarrow RO \rightarrow HO_2 \rightarrow OH$ radical point involving the radicals originated in the degradation of VOCs. The entire process, which also includes reactions that inhibit ozone formation, is peculiar to each VOC formula due to the different relative lifetimes, oxidation rates, sizes, and structures of the radicals generated.

Therefore, as seen in Eq. (3.2), POCPi was used as a characterization factor. This is defined as the quotient between the extra amount of ozone formed by

BOX 3.2 Calculation application example for photochemical ozone creation potential category—cont'd

VOC emissions and ethylene, which is taken as a reference and whose POCP is 100, following Eq. (3.2) (Andersson-Sköld & Holmberg, 2000).

Through this equation, the POCPi values can be calculated for each individual VOC. However, Annex 7 (Table A.7) of the attached material presents the main characterization factors following the databases reported in the literature (Altenstedt & Pleijel, 1998; Derwent, Jenkin, & Saunders, 1996; Derwent, Jenkin, Passant, & Pilling, 2007).

At the exemplification level, this impact category can be used to monitor the effectiveness of VOC removal from an industrial biofilter located in a waste management plant (WMP). Some sample data are considered as the amount of mass inlet concentration of the volatile compound of 1.28 kg butane/m^3. After verifying that your POCPi has an attribution of 35.2 C_2H_4/kg eq, the ozone creation potential can be quantified by multiplying the mass by the characterization factor:

$$POCP = \sum POCP_i \times m_i$$

$$POCP = 35.2\,\text{kg}\,C_2H_{4eq} \times 1.28\frac{\text{kg}}{\text{m}^3}$$

$$POCP = 45.05\,\text{kg}\,C_2H_{4eq}/\text{m}^3$$

Therefore, the estimated photochemical ozone creation potential for this process is 45.05 kg C_2H_{4eq}/m^3.

Ecotoxicity

Ecotoxicity potential is quantitative substance specific of potential impacts per unit of emission of a toxic substance that can be used as weighting factors in the aggregation of emissions from life cycle inventories.

From this concept, classifications were established according to the degree of harmfulness resulting from interactions of chemical substances with organisms. Thus, the tests developed were divided into acute toxicity and chronic toxicity. These tests basically differ in duration and in the final responses that are measured.

Acute toxicity tests are used to measure the effects of toxic agents on aquatic species over a short period of time over the life span of the test organism. They are intended to estimate the dose or concentration of a toxic agent that would be capable of producing a measurable specific response in a test organism or population in a relatively short time, generally 24–96 hours. Generally, the values of effective and lethal concentrations are expressed in relation to 50% of organisms because these responses are more reproducible, can be estimated with a higher degree of reliability, and are more significant to be extrapolated to a population. In the aquatic environment, acute effects caused by toxic agents on organisms may result from inappropriate applications of pesticides, from environmental accidents, and from situations where untreated industrial effluents are released into receiving water bodies.

In contrast, chronic toxicity tests are performed to measure the effects of chemicals on aquatic species for a period that may cover part or all of the test organism's life cycle. The fact that a chemical does not produce toxic effects on aquatic organisms in acute toxicity tests does not indicate that it is not toxic to them. Moreover, they allow to assess the possible toxic effects of chemicals under conditions of prolonged exposures to sublethal concentrations, i.e., concentrations that allow the survival of organisms but which affect their biological functions, such as reproduction and development, growth and maturation, among others.

Additionally, considering the harmfulness of the toxicity tests, differentiations regarding the place of occurrence of this contamination have been determined. Thus, within the potential for ecotoxicity, it is possible to characterize freshwater aquatic ecotoxicity potential (FAEP), marine aquatic ecotoxicity potential (MAEP), and terrestrial ecotoxicity potential (TETP). In short, these potentials present the response of the emission of some substances referring to the maximum tolerable concentrations for the environment, whether they are aquatic of freshwater, marine, or terrestrial.

As in every impact category, characterization factors are used to estimate the values of impacting masses. The reference units used by the impact category of the toxicity potential are commonly expressed in kg 1,4-dichlorobenzene equivalent (1,4-DB). However, characterization factors are currently used as extrapolations of the same substance from aquatic systems to terrestrial systems. In this sense, characterization factors for ecotoxicity are obtained according to Vaal and Folkerts (1998) with Eq. (3.3):

$$ETP_{i,ecomp} = \frac{\sum_{fcom} F_{i,ecomp} \times E_{i,ecomp}}{\sum_{fcom} F_{ref,ecomp,fcom} \times E_{ref,fcomp}} \tag{3.3}$$

where $ETP_{i,ecomp}$ is the toxicity potential when the substance i is released into the environmental medium *ecomp*. $F_{i,ecomp}$ is the amount of the released i that reaches *fcomp*, the final environmental medium just before exposure, *fcom* of F_i is the fate factor, $E_{i,fcomp}$ is the toxic effect of the exposure of the organisms growing or living in *fcomp* to *i,fcom* of E_i is the effect factor, and *ref* is the base reference substance.

Thus, after defining the factor of characterization, Eq. (3.4) is used to quantify the ecotoxicity potential.

$$ETP_{(FARP,\, MAEP,\, TEP)} = \sum_i ETP_i \times m_i \tag{3.4}$$

where *ETP* is the ecotoxicity potential corresponding to the score of *freshwater aquatic ecotoxicity potential (FAEP), marine aquatic ecotoxicity potential*

(*MAEP*), and *terrestrial ecotoxicity potential* (*TETP*), ETP_i is the characterization factor for the substance *i* emitted to the environmental (freshwater, marine, or terrestrial) (kg eq), and m_i is the amount (in mass units) emitted to a given environmental compartment (kg), expressed in kg 1,4-DB-eq.

Therefore, a brief ecotoxicity potential category outline can be seen in Table 3.5, and a calculation exemplification is inserted in Box 3.3.

TABLE 3.5 Ecotoxicity category outline.

Impact category	Ecotoxicity
Definition	Toxic effects of chemical on an ecosystem
Impact indicator	Biodiversity loss and extinction of species
Considerations	Toxicological responses of different species and nature of the chemical in the ecosystem
Damage categories	Ecosystem quality and species extinction
Unit	kg 1,4-DB equivalent or PDF (Potentially disappeared fraction of species) or PAF (Potentially affected fraction of species)

BOX 3.3 Calculation application example for ecotoxicity potential category.

In an LCA, a toxicity impact score can be calculated using Eq. (3.4).

For example, emissions of benzo[α]pyrene of 0.1 kg for continental air and 0.2 kg for continental freshwater (per functional unit of m^3). However, note that the characterization factors for the same compound differ in terms of emissions to air or freshwater. Thus, we will use the factors of 3.01×10^{-6} and 1.26×10^{-5} CTUe for air and freshwater, respectively. Therefore, the sum of the potential for human toxicity is given by:

$$ETP_{(air,\ freshwater)} = \sum_{benzo-\alpha-pyrene} (3.01 \times 10^{-6}\,CTU_e \times 0.1\,kg) + (1.26 \times 10^{-5}\,CTU_e \times 0.2\,kg)$$

$$ETP_{(air,\ freshwater)} = 2.82 \times 10^{-6}\,CTU_e$$

This human toxicity impact score can then be added to those of other substances in the inventory. The toxicity factors, i.e., the characterization factors, presented here, should be used in a way that reflects the wide variation often in orders of magnitude uncertainty between the chemical characterization factors, as well as the three orders of magnitude uncertainty about the individual factors. In addition, each impact factor must be consulted in the literature or calculated according to the place of issue and the legislation indicators.

Acidification potential

Acidification is a phenomenon resulting from the release of protons from certain substances into the ecosystem. These emissions increase the acidity (decrease in pH) of water and soils, giving rise to the most frequent environmental problem being acid rain (Reuss & Johnson, 2012).

Acidification potential (AP) can be divided into two classes, known as impacts on terrestrial ecosystems or aquatic ecosystems. Basically, their consequences for terrestrial ecosystems are related to the inefficient growth of forests or their devastation, while in aquatic ecosystems their impacts imply clean acid lakes, that is, with no observed wildlife.

Among the main substances that contribute to acidification are substances that result in the supply or release of hydrogen ions (H^+) and the leaching of the anions corresponding to the system to be analyzed. In particular, they include sulfur dioxide (SO_2), sulfur trioxide (SO_3), nitrogen oxides (NO_x), hydrogen chloride (HCl), nitric acid (HNO_3), sulfuric acid (H_2SO_4), phosphoric acid (H_3PO_4), hydrogen fluoride (HF), hydrogen sulfide (H_2S), and ammonia (NH_3). The deposition of acidifying chemicals can be direct or via precipitation and the impact depends on the sensitivity of the receiving medium. Therefore, the specific characterization of acidification often produces significantly different results for each characterization factor (Table 3.6).

Also, since SO_2 is used as the basis for the determination of acidification potential or equivalence factor, the method of establishing effect factors for acidification substances is based on stoichiometric considerations and is internationally accepted. Equivalence factors are determined by Eq. (3.5) (Hauschild & Wenzel, 1998).

$$EF = \frac{2}{2 \times M_W} \times 64.06 \quad (3.5)$$

TABLE 3.6 SO_2 equivalence factors of various acid producers.

Acid producer (in air)	SO_2 equivalence factor
1 kg HCl	0.88 kgeq SO_2
1 kg HF	1.60 kgeq SO_2
1 kg NO_2	0.70 kgeq SO_2
1 kg SO_2	1.00 kgeq SO_2
1 kg H_2S	1.88 kgeq SO_2
1 kg NH_4	0.89 kgeq SO_2
1 kg NH_3	0.93 kgeq SO_2

Based on ILDC (2010).

where EF is the equivalence factor for substance, M_W is the molecular weight of the substance emitted, n is the number of hydrogen ions released in the recipient as a result of conversion of the substance, and 64.06 g/mole is the molecular weight of SO_2.

Acidification potential is quantified by using acidification potential (AP) for substances having the same effect as SO_2 in reflection of acidification (Eq. 3.6). Acidification potentials are expressed as SO_2-equivalents (kg SO_2-eq) or mol H^+.

$$AP = \sum_i AP_i \times m_i \tag{3.6}$$

where AP is the acidification potential, AP_i is the equivalence factor for the substance i (kg eq), and m_i is the emission of substance i (kg), expressed in kg SO_2-eq.

Therefore, a brief acidification potential category outline can be seen in Table 3.7, and a calculation exemplification is inserted in Box 3.4.

TABLE 3.7 Acidification category outline.

Impact category	Acidification
Definition	Reduction of the pH due to the acidifying effects of anthropogenic emissions
Impact indicator	Increase of the acidity in water and soil systems
Considerations	Acidifying potential of oxides of nitrogen and sulfur
Damage categories	Ecosystem quality and decrease in biodiversity
Unit	kg SO_2 equivalent or mol H^+

BOX 3.4 Calculation application example for acidification potential category.

Hydrogen sulfide (H_2S) is known to be a corrosive, poisonous, and gaseous compound in its natural state. This compound is generally found on energy platforms, mainly as a volatile residue from coal synthetic gas, natural gas, and in types of oil that contain sulfur. Thus, to quantify the acidification potential, the first step is to determine the equivalence factor. This is given through Eq. (3.4), where the molecular weight of the substance emitted is quantified and the number of hydrogen ions released in the recipient as a result of the conversion of the substance (hydrogen sulfide has two molecules):

$$EF = \frac{n}{2 \times M_W} \times 64.06$$

Continued

BOX 3.4 Calculation application example for acidification potential category—cont'd

$$EF = \frac{2}{2 \times 34.1} \times 64.06$$

$$EF = 1.88\,\text{kg}\,SO_2\,\text{eq}$$

In terms of exemplification, hydrogen sulfide, in a concentration range of 10–50ppm, already causes symptoms such as headache, nausea, throat, and irritated eyes, resulting in even eye damage. In this way, we have considered the maximum concentration of 50ppm, i.e., 0.050kg/m^3. However, to assess the potential environmental damage caused by this gas, it is necessary to apply the equivalence factor.

$$AP = \sum_i AP_i \times m_i$$

$$AP = 1.88\,\text{kg}\,SO_2\,\text{eq} \times \frac{0.050\,\text{kg}}{\text{m}^3}$$

$$AP = 9.4 \times 10^{-2}\,\text{kg}\,SO_2\,\text{eq/m3}$$

Therefore, the acidification potential caused by hydrogen sulfide represents about 9.4×10^{-2} kg SO_2 eq.

Eutrophication potential

The category of eutrophication potential (EP) can be found under various nomenclatures, among them nutrient enrichment or nitrification. This category aims to characterize the impacts of nutrients such as nitrogen and phosphorus in their bioavailable forms on ecosystems. In addition, as well as the potential for acidification, the potential for eutrophication also has ways of classifying the systems to be analyzed as terrestrial ecosystems and aquatic ecosystems.

In terrestrial systems, nutrient uptake can modify species of vegetation composition, favoring species that prioritize higher levels of nutrients to grow more rapidly than more nutrient-efficient plants. The terrestrial impact is directly related to the deposition of emissions of nitrogen compounds as nitrogen oxides from combustion, and agricultural ammonia. It is important to emphasize that the propagation of phosphorus in the air is not considered influential terrestrial eutrophication.

At the same time, in aquatic systems, nutrient enrichment has a similar impact on the terrestrial ecosystem. Thus, the composition of aquatic species, such as algae, favors the available nutrients, multiplying rapidly, causing shadows in the place, and changing the living conditions of both the algae themselves that require the light to perform the photosynthesis and the fish that need light to capture their fighters. Also, in the face of these consequences, the algae in decomposing cause oxygen depletion in the bottom of the body of water, that is, the dissolved oxygen in the medium is consumed with greater facility through the aerobic decomposition. Therefore, it can lead to serious damage to the biological populations that inhabit the sediment. In addition, direct toxic effects on higher organisms, including humans, should be considered when certain species of algae appear in mass.

Additionally, within the aquatic systems, there is the subdivision of freshwater systems and marine systems. While the first has as a limitation the nutrient phosphorus, in marine systems the limiting nutrient is nitrogen. However, there may be variations depending on the locality of the systems as well as exposures to atmospheric emissions.

Since marine, as well as terrestrial systems, have nitrogen as the limiting factor, phosphorus is determined as the degree of autotrophic activity in most cases. However, there are substantial suggestions to include potencies of eutrophication such as chemical oxygen demand (COD), biological oxygen demand (BOD), and nitrates (NO_3^-) in the form of phosphate (PO_4). Therefore, Table 3.8 presents the main equivalence factors suggested by the website of the Institute of Environmental Sciences (CML) and generally proposed by the LCA.

Given these aspects, the characterization of the eutrophication potential can be measured according to Eq. (3.7).

$$EP = \sum_i EP_i \times m_i \tag{3.7}$$

where EP is the eutrophication potential, EP_i is the equivalence factor for the substance i (kg eq), and m_i is the emission of the substance i (kg), expressed in kg PO_4-eq.

Therefore, a brief eutrophication potential category outline can be seen in Table 3.9, and a calculation exemplification is inserted in Box 3.5.

TABLE 3.8 PO_4 **equivalence factors of various substances.**

Nutrient	PO_4 equivalence factor
1 kg Nitrogen oxides (NO_x, air)	0.13 kg eq PO_4
1 kg Total nitrogen (water)	0.42 kg eq PO_4
1 kg Total phosphorus (water)	3.07 kg eq PO_4
1 kg Chemical oxygen demand (COD)	0.022 kg eq PO_4
1 kg NH_3	0.35 kg eq PO_4
1 kg NH_4^+	0.33 kg eq PO_4
1 kg NO_3^-	0.095 kg eq PO_4
1 kg NO_2^-	0.13 kg eq PO_4

Based on ILDC (2010).

TABLE 3.9 Eutrophication category outline.

Impact category	Eutrophication
Definition	Accumulation of nutrients in aquatic systems
Impact indicator	Increase of nitrogen and phosphorus concentrations, and formation of biomass (e.g., algae)
Considerations	Transportation of nutrients (air, water, wash-off from land)
Damage categories	Ecosystem quality
Unit	kg PO_4 equivalent (in limbic area) or kg N equivalent (in marine and terrestrial ecosystems)

BOX 3.5 Calculation application example for eutrophication potential category.

As can be seen, the category of eutrophication potential is mainly attributed to the enrichment of nutrients in freshwater or marine aquatic surfaces, as well as the land surface.

It is widely known that the increase in nitrogen (N) and phosphorus (P) fertilizers shows a significant improvement in relation to the performance of agro-food production systems. However, as a result of this excessive demand, a significant fraction of N and P applied to agricultural land is transported by rivers to coastal areas, resulting in water eutrophication.

In monoculture systems, it is estimated that the emissions generated by the application of fertilizers are responsible for including losses of total N, NO_3^-, NH_4^+, the soluble phosphate, and total P, by flow and leaching. Thus, at the level of exemplification, in rice monocultures, the application rate of about 171.75 kg/ha of a nitrogen-based fertilizer (NO_3^-) was determined. However, its runoff rate (net loss) represents about 0.65 kg/ha. Thus, when we consult Table 3.8, we find that the NO_3^- characterization factor is 0.095 kgeq PO_4 (see Table 3.8). Therefore, the eutrophication potential can be calculated as:

$$EP = \sum_i EP_i \times m_i$$

$$EP = \sum_i 0.095\ \text{kg}\ PO_4\ \text{eq} \times 0.65\ \text{kg}$$

$$EP = 6.1 \times 10^{-2}\ \text{kg}\ PO_4\ \text{eq}$$

Therefore, the eutrophication potential can be attributed to rice fields of approximately 6.1×10^{-2} kg PO_4 eq per hectare.

Energy assessments

Increasing energy consumption has become a major challenge worldwide. Concerns about the depletion of fossil resources have fueled the search for environmentally sustainable alternative sources to meet the energy demands.

It is known, however, that energy analyses can be calculated to complement the life cycle analysis studies of a production system. The life cycle energy analysis tool (LCEA) was proposed and developed to present a more simplified and direct approach to the energy relationships of the processes. Thus, this procedure assesses the performance of life cycle inventories, directly assigning energy coefficients to the inflows and outflows.

Additionally, the LCEA has also been applied as a study parameter in LCA topics, informing the energy efficiency of a production process, providing indicators of net energy balance or energy ratio. Given this perspective, several methodologies for calculating the energy efficiency of production systems are found in the literature. However, the use of such methodologies is sometimes ambiguous and can lead to results that are somewhat difficult to compare (Mourad & Walter, 2011).

The LCEA is basically applied by converting the data collected in the life cycle inventory into energy flows. Thus, this parameter allows the establishment of a logical structure to quantify the net energy of different inflows and outflows, as well as to calculate the energy efficiency of the processes. However, depending on the methodology used, the results may provide different information. Therefore, it is important to emphasize that the results obtained in the energy evaluation of the production systems can provide information about the net energy obtained, in units of total energy or specific energy (that is, energy/year; energy/area; etc.), as well as inform the (dimensionless) relationship between the energy obtained and the energy used in the life cycle.

On the whole, Eqs. (3.8) and (3.9) state the types of LCEA indicators.

$$NE = E_{out} - E_{in} \tag{3.8}$$

$$ER = \frac{E_{out}}{E_{in}} \tag{3.9}$$

where NE is the net energy, ER is the energy ratio, E_{out} is the energy output, and E_{in} is the energy input. Furthermore, energy flows may or may not consider renewable energy inputs. All units can be energy/year.

However, to categorize the energy indicators, it is possible to separate into two major divisions, being the liquid energy indicators and the energy ratio indicators. The liquid energy indicators usually express the differences between the incoming and outgoing energy flows within a process. Currently, it is estimated that about five parameters are used: they are net energy gain, net energy balance, net energy yield, net energy production, and net energy value.

The indicator of net energy gain (NEG) is generally calculated in the activities of biomass and bioenergy production, evaluating the production systems in terms of their contribution to defining renewable energy goals. However, this equation does not clearly indicate whether the allocation can be used, that is, whether the indicator accounted for only the energy contained in the product or also in the coproducts, as well as whether the energy inputs are renewable and/or nonrenewable. Thus, the NEG can be calculated using Eq. (3.10), as described by Arodudu, Ibrahim, Voinov, and van Duren (2014):

$$NEG = E_{out} - E_{in} \tag{3.10}$$

where NEG is the net energy gain (MJ/m), E_{out} is the total energy produced by the system (MJ/m), and E_{in} is the total energy used in the system (MJ/m).

In addition, the net energy balance (NEB) can be determined by Eq. (3.11), where the total values of the fossil energy inputs used to generate the product are subtracted from the total energy values of the products and their byproducts (Hill, Nelson, Tilman, Polasky, & Tiffany, 2006).

$$NEB = E_{out} - E_{in} \tag{3.11}$$

where NEB is the net energy balance (MJ/L), E_{out} is the total energy produced by the system (MJ/L), and E_{in} is the total energy used in the system (MJ/L). Note that the units can be energy/mass or energy/volume.

The net energy yield (NEY) is suggested as the most appropriate net energy metric when the objectives concern the displacement of imported energy and the limited availability of land, being considered a combination of efficiency and productivity. Thus, as defined in Eq. (3.12), the NEY is simply the production of gross energy minus the entry of gross energy (Fore, Porter, & Lazarus, 2011).

$$NEY = E_{out} - E_{in} \tag{3.12}$$

where NEB is the net energy yield (MJ/ha), E_{out} is the total energy produced by the system (MJ/ha), and E_{in} is the total energy used in the system (MJ/ha).

The net energy production (NEP) can be calculated using Eq. (3.13), according to the methodology proposed by Kamahara et al. (2010):

$$NEP = (EF + EB) - EI \tag{3.13}$$

where NEP is the net energy production (GJ/ha), EF is the energy output of the fuel (GJ/ha), EB is the energy output of byproducts (GJ/ha), and EI is the total energy inputs (GJ/ha).

As with the other energy indicators, the net energy value (NEV) is characterized by being highly sensitive to the assumptions about the system limits and the values of the key parameters and, regarding the energy content, since it considers the areas required in the productive systems (Shapouri, Duffield, & Wang, 2002). This indicator is defined as the energy content of the product minus fossil energy used to produce the same (Eq. 3.14).

$$NEV = E_{out} - E_{in} \tag{3.14}$$

where *NEV* is the net energy value (MJ/ha), E_{out} is the total energy produced by the system (MJ/ha), and E_{in} is the total energy used in the system (MJ/ha).

On the other hand, the energy ratio indicators used usually express the relationships between the input and output energy flows within a process. Currently, it is estimated that about seven parameters are used: energy return on energy invested, net energy ratio, renewable factor, primary energy balance, the energy renewability efficiency, the life cycle energy efficiency, and fossil energy ratio.

The energy return on energy invested (EROI) indicator, proposed by Hall (1972), can be considered an indicator of the overall efficiency of the process. EROI is often used to express the advantages and disadvantages of various fuels, considering whether they are renewable or not. EROI is defined in Eq. (3.15), as provided by Arodudu et al. (2014):

$$EROI = \frac{OE}{IE} \tag{3.15}$$

where *EROI* is the energy return on energy invested (dimensionless), *OE* is the output energy (MJ), and *IE* is the input energy (MJ).

The net energy ratio (NER) is a model widely used in the literature to evaluate the energy performance of industrial products and agricultural processes. Its purpose is to relate the part of total fossil energy required by the processes, with the absorbed energy available in the final products. In addition, the NER has been widely used to address the efficiency of technologies in terms of energy supply to society.

In parallel, one of the main bottlenecks for the development of production systems on a commercial scale is related to the difficulty of obtaining a positive net energy relationship. However, greater energy efficiency is a crucial factor when production systems are designed. In general, the net energy ratio (NER) is applied to estimate the sustainability and efficiency of the system, indicating whether the production process results in energy loss or gain. Thus, the higher the NER, the more efficient the energy system will be. In addition, some authors suggest that NER should be greater than 1 (Farid et al., 2020; Sun et al., 2019). However, most scheduling processes do not meet this theoretically established condition, indicating that the NER metric should be improved through optimization. Therefore, to quantify this energy relation is used Eq. (3.16) (Jorquera, Kiperstok, Sales, Embirucu, & Ghirardi, 2010):

$$NER = \frac{\sum E_{out}}{\sum E_{in}} \tag{3.16}$$

where *NER* is the net energy ratio (dimensionless), E_{out} is the energy produced (MJ), and E_{in} is the total input energy required (MJ).

Additionally, the Renewable Factor (RF) aims to express how independent the fossil energy system is. In the literature, although the NER and RF have similar characteristics, and are sometimes related, some differences can be noted. According to Pradhan, Shrestha, Van Gerpen, and Duffield (2008), the NER can be applied as an energy efficiency indicator, while the RF as a renewable factor. In this way, the RF can be calculated using Eq. (3.17):

$$RF = \frac{\sum E_{out}}{\sum NR_{E_{in}}} \tag{3.17}$$

where RF is the renewable factor (dimensionless), E_{out} is the energy output (MJ), and $NR_{E_{in}}$ is the nonrenewable energy input (MJ).

Primary energy balances (PEBs) can be expressed as the ratio between primary fossil energy inputs to renewable energy output (García et al., 2011). This indicator notoriously addresses that the energy inputs are of exclusively fossil origin and the energy output is focused on a certain product, as they make use of allocation criteria (Eq. 3.18).

$$PEB = \frac{\sum RE_{out}}{\sum FE_{in}} \tag{3.18}$$

where PEB is the primary energy balance (dimensionless), RE_{out} is the renewable energy output (MJ), and FE_{in} is the fossil energy input (MJ).

The efficiency of energy renewability (ERenEf) is an indicator proposed by Malça and Freire (2006), to characterize the renewability of an energy source. The ERenEf measures the fraction of the final energy of the product obtained from renewable sources. In addition, this indicator can be used with or without allocation criteria, being calculated using Eq. (3.19):

$$ERenRf = \frac{(FEC - Ein, fossil, prim) \times 100}{FEC} \tag{3.19}$$

where $ERenEf$ is the energy renewability efficiency (dimensionless), FEC is the fuel energy content (MJ), and $Ein,fossil,prim$ is the input required (MJ), measured in terms of primary energy.

In addition, it is noteworthy that, a product can be considered renewable if the ERenEf assumes values between 0% and 100%. If there were no nonrenewable energy inputs, the product would be fully renewable with an ERenEf of 100%. If the ERenEf is less than zero, then the product must be characterized as nonrenewable, since the nonrenewable energy needed to grow and convert raw material into a product would be greater than the energy present in the final product. In this case, by definition, nonrenewable energy sources have negative ERenEf values, with negative values increasing as the energy efficiency of the life cycle decreases. Therefore, if $0 < \text{ERenEf} < 100$, the process is renewable, and if $\text{ERenEf} < 0$, the process is not renewable.

According to Sheehan et al. (1998), there are two ways to assess energy efficiency. For this, the first form is attributed to Eq. (3.20), where an indicator of the life cycle energy efficiency (LCEE) is proposed. The LCEE can be quantified simply by the relationship between the energy of the product and the total primary energy. This efficiency is responsible for the energy losses of the raw material and additional process energy needed to make the fuel. In contrast, the second model consists of the fossil energy ratio (FER) (Eq. 3.21). This, in turn, tells us something about the degree to which a given product is renewable or not. Therefore, it is defined simply as the ratio between the energy of the final product and the amount of fossil energy needed to do so.

$$LCEE = \frac{FPE}{TPE} \tag{3.20}$$

$$FER = \frac{FE}{FE_{in}} \tag{3.21}$$

where *LCEE* is the life cycle energy efficiency (dimensionless), *FPE* is the fuel product energy (MJ), and *TPF* is the total primary energy (MJ). The *FER* is the fossil energy ratio (dimensionless), *FE* is the fuel energy (MJ), and *FEI* is the fossil energy inputs (MJ).

Additionally, it is important to note that if the proportion of fossil energy has a zero value, a product is not only completely nonrenewable but also does not provide usable product energy as a result of the fossil energy consumed to make the product. If the proportion of fossil energy is equal to 1, then this product is not yet renewable. A fossil energy ratio of 1 means that no energy loss occurs in the process of converting fossil energy into a usable product. For fossil energy rates greater than 1, the product actually starts to provide leverage of fossil energy. As a product approaches being completely renewable, the proportion of fossil energy approaches infinite value. In other words, a fully renewable product has no fossil energy requirements.

Furthermore, according to Mayer, Brondani, Carrillo, Hoffmann, and Lora (2020), a standardization model for energy efficiency indicators was proposed (Table 3.10). Table 3.10 presents a summary of the energy efficiency indicators and the similarity regarding the use of the equations applied to the allocation criteria, type of energy input, and considering the use of coproducts. Through this compilation of information, it is possible to observe that the equations for calculating the EROI, NER, and LCEE present several similarities in terms of application since all four consider the overall energy efficiency and the sustainability of the system. Finally, depending on which components were assumed, the energy efficiency indicators can be related to other equations, if they make use of allocation and consider only nonrenewable energy as input. Therefore, it is pertinent that the equations to calculate EROI, NER, ER, and LCEE consider both renewable and nonrenewable energies as inputs to the system to indicate the overall efficiency and sustainability of the evaluated system.

TABLE 3.10 Energy efficiency indicators and their characteristics for energy assessments in production systems.

Indicator	Allocation criteria	Inputs flow		Energy of products considered in equation	Indicator characteristic
		Renewable	*Nonrenewable*		
NEG	✓[a]	✓	✓	✓[b]	Net energy and renewability
NEY	✓[a]	✓	✓	✓[b]	Net energy and renewability
NEB	✓[a]	✓	✓	✓[b]	Net energy and renewability
NEP	✓[a]	✓	✓	✓[b]	Net energy and renewability
NEV	✓[a]	✓	✓	✓[b]	Overall efficiency and sustainability
EROI	✓[a]	✓	✓	✓[b]	Overall efficiency and sustainability
NER	✓[a]	✓	✓	✓[b]	Overall efficiency and sustainability
ER	✓[a]	✓	✓	✓[b]	Overall efficiency and sustainability
LCEE	✓[a]	✓	✓	✓[b]	Overall efficiency and sustainability
FER	✓	χ	✓	χ	Renewability
ERenEf	✓	χ	✓	χ	Renewability
RF	✓	χ	✓	χ	Renewability
PEB	✓	χ	✓	χ	Renewability

Note: The icons correspond to yes (✓) and no (χ).[a] means that these indicators may not consider allocation criteria; [b] means that these indicators may not consider energy from coproducts.
Based on Mayer et al. (2020).

Therefore, a brief energy assessments category outline can be seen in Table 3.11, and a calculation exemplification in inserted in Box 3.6.

TABLE 3.11 Energy assessments category outline.

Impact category	Energy assessments (depletion of abiotic resources)
Definition	Decrease in the availability of nonbiological resources (non- and renewable) as a result of their unsustainable use
Impact indicator	Decrease in resources
Considerations	Distinctions between renewable and nonrenewable resources
Damage categories	Ecosystem quality (natural resources and possible ecosystem collapse)
Unit	Dimensionless, units of energy/mass, energy/year, or energy/hectare

BOX 3.6 Calculation application example for energy assessments category.

The relentless search for the reduction of the presence of fossil fuels in the energy matrixes of industrial and agro-industrial processing has become, by far, one of the most important strategies for the conservation and preservation of the environment, aiming at sustainable development.

As mentioned throughout the topic of energy assessments, the methodologies reported in the literature basically comprise liquid energy indicators and energy indexes. These, in turn, result in equations that need to consider particularities of each limit of the specific system as allocation criteria, characteristics of inputs, as well as the characteristic of energy origin, be it fossil or renewable.

Therefore, we will exemplify the most used equations to determine the energy assessment indicators for life cycle assessments.

Net energy and energy ratio equations are generic models. This means that energy flows may or may not consider renewable energy inputs. For example, if we evaluate a petroleum coke combustion system, where the input of fossil energy requires 12 MJ, the output of the system has a fossil energy potential of 5 MJ.

When we apply to determine the net energy and energy ratio we will have:

$$NE = E_{out} - E_{in}$$

$$NE = 5\,\text{MJ} - 12\,\text{MJ}$$

$$NE = -7$$

$$ER = \frac{E_{out}}{E_{in}}$$

Continued

BOX 3.6 Calculation application example for energy assessments category—cont'd

$$ER = \frac{5\,\text{MJ}}{12\,\text{MJ}}$$

$$ER = 0.41$$

Therefore, the net energy and energy ratio will be 7 and 0.41, respectively. It is important to note that energy inputs were considered fossil energies. If we wanted to consider renewable energy, it would be possible and at the discretion of the evaluator. However, if we used renewable energies, the most appropriate would be to use the following equations:

$$NEB = E_{out} - E_{in}$$

$$NER = \frac{\sum E_{out}}{\sum E_{in}}$$

For example, if we evaluated an energy biomass production system, whose fossil energy inputs consist of 48.34 MJ/d and the renewable energy output is around 31.21 MJ/d, so:

$$NEB = 31.21\,\frac{\text{MJ}}{\text{d}} - 48.34\,\frac{\text{MJ}}{\text{d}}$$

$$NEB = -17.13$$

$$NER = \frac{\sum 31.21\,\frac{\text{MJ}}{\text{d}}}{\sum 48.34\,\frac{\text{MJ}}{\text{d}}}$$

$$NER = 0.64$$

Therefore, the NEB and NER results would be 17.13 and 0.64, respectively.

On the other hand, net energy gain (NEG) determinations are generally applied to assess the energy gain or loss of spent fuels, which means that a high NEG value refers to high energy efficiency. Thus, when assessing the life cycle of ethanol production and the energy content of ethanol (21.1 MJ per L), it was determined that the fossil energy input had a demand of 15.24 MJ/L. When applying the values in the equation we will have an energy gain of 5.86, as follows:

$$NEG = E_{out} - E_{in}$$

$$NEG = 21.1\,\frac{\text{MJ}}{\text{L}} - 15.24\,\frac{\text{MJ}}{\text{L}}$$

$$LNEG = 5.86$$

Additionally, the net energy indicators can be given through the NEY and NEV determinations. Both indicators are used to quantify the production of renewable energy generated from the production of bioenergy.

$$NEY = E_{out} - E_{in}$$

$$NEV = E_{out} - E_{in}$$

Therefore, when evaluating the net energy of an ethanol production process from corn, the input values of the process were assumed to be 250.18 MJ/ha and the

BOX 3.6 Calculation application example for energy assessments category—cont'd

output presented around 335.20 MJ/ha. Therefore, both NEY and NEV can represent around 85.02.

$$NEY = 335.20\frac{\text{MJ}}{\text{ha}} - 250.18\frac{\text{MJ}}{\text{ha}}$$

$$NEY = 85.02$$

In contrast, the net energy production (NEP) can be calculated using the following equation:

$$NEP = (EF + EB) - EI$$

At the level of exemplification, when determining the liquid energy production of an industrial process where there is the generation of 1 kg of biomass and 1 L of oil as fuel, both with an energy potential of 15.2 MJ/kg and 23.12 MJ/L, respectively, a fossil energy input of 52.15 MJ/kg of biomass is required:

$$NEP = (23.12 + 15.2) - 52.15$$

$$NEP = -13.83$$

Therefore, the process had a net negative energy production of 13.83.

In addition, efficiency indicators such as energy return on energy invested (EROI) can assist in the overall determination of process efficiency. However, this methodology does not specify whether the energy flow is from renewable or fossil sources. However, if we considered renewable energy, as a study objective, we should choose the primary energy balance (PEB) indicator. This, in turn, determines the renewable energy of the process.

$$EROI = \frac{OE}{IE}$$

$$PEB = \frac{\sum RE_{out}}{\sum FE_{in}}$$

Thus, let us consider a hypothetical scenario, where the energy input and output of the process represent about 185.28 and 197.14 MJ, respectively. However, we have not specified whether it is renewable energy or not. Therefore, we would apply the EROI methodology:

$$EROI = \frac{197.14\,\text{MJ}}{185.28\,\text{MJ}}$$

$$EROI = 1.06$$

Therefore, we would have an invested return energy of 1.06. However, if we specify that the output energy is renewable energy, considering a bioenergy process, we would determine the primary energy balance of 1.06.

In parallel, some energy indicators have been applied exclusively to bioenergy processes. Among some examples, the renewability factors (RF), the efficiency of energy renewability (ERenEf), the life cycle energy efficiency (LCEE), and fossil energy ratio (FER). However, indicators such as RF, LCEE, and FER can be applied for the same purpose.

Continued

BOX 3.6 Calculation application example for energy assessments category—cont'd

Thus, at the level of exemplification, renewable energy factors are generally applied to evaluate new biofuel proposals, such as microalgae biodiesel. To this end, considering that the energy input for the production of microalgae biodiesel requires about 130 MJ, and the bioenergetic potential of the fuel is approximately 23 MJ:

$$RF = \frac{\sum E_{out}}{\sum NR_{E_{in}}}$$

$$RF = \frac{23\,\text{MJ}}{130\,\text{MJ}}$$

$$RF = 0.17$$

Therefore, the microalgae biofuel renewability factor is 0.17.

In addition, we can evaluate the efficiency of the energy renewability indicator for the same scenario:

$$ERenRf = \frac{(FEC - Ein,\ fossil,\ prim) \times 100}{FEC}$$

$$ERenRf = \frac{(23\,\text{MJ} - 130\,\text{MJ}) \times 100}{23\,\text{MJ}}$$

$$ERenRf = -465.21$$

Therefore, the renewability index resulted in a negative value of 465.21.

Water footprint

The difficulty of quantifying the different uses of water used in the products or processes, methodologies such as the evaluation of the water footprint were developed with the purpose of instructing institutions to follow a sequence of steps in the search for the correct evaluation of the impacts caused by certain resources.

However, only recently have water footprint methods aligned with life cycle assessment approaches. Thus, a significant step along this path was the development of the ISO 14046 standard for the water footprint (ISO, 2014), which has more explicitly adhered to the water footprint within the life cycle assessment methodology.

In the past, the water footprint was estimated only on the blue footprint, i.e., the volume of surface (or subsurface) water, evaporated or incorporated into the products. However, currently, the main renewal of the water footprint is directly related to the division of water resources into three categories: blue, green, and gray (Mekonnen & Hoekstra, 2011). Thus, besides the definition of the water footprint blue, the green footprint can be defined as the volume of rainwater evaporated or absorbed by the product (usually quantified in agricultural production systems). Yet, the gray footprint is defined as the volume of water needed to dilute the pollution generated in the production of a given product or process to the levels established by current environmental standards.

Additionally, a further deepening of water consumption can be subdivided into water consumption in the supply chain (indirect) and consumption within the production chain (direct). This classification allows a broader perspective on the origin of consumption and facilitates the choice of suppliers that present an efficient use of the natural resource or the reduction of consumption in the production chain (Fig. 3.2).

The total water footprint can be quantified according to Eq. (3.22) described by Hoekstra and Mekonnen (2012) and defined as the sum total of its blue, green, and gray footprints considering direct or indirect consumption.

$$WF = \sum WF_{blue} + WF_{green} + WF_{gray} \quad (3.22)$$

where WF is the total water footprint, WF_{blue} is the blue water footprint (m^3), WF_{green} is the green water footprint (m^3), and WF_{gray} is the gray water footprint (m^3), total expressed in m^3.

Moreover, the blue water footprint is an indicator of the consumptive use of the so-called blue water, that is, fresh surface or groundwater. The steps of blue footprints can be quantified according to Eq. (3.23):

$$WF_{blue} = BW_e + BW_i + L_{rf} \quad (3.23)$$

where WF_{blue} is the blue water footprint, BW_e is blue water evaporation (m^3), BW_i is blue water incorporation (m^3), and L_{rf} is the lost return flow (m^3), total expressed in m^3.

In contrast, the green water footprint is the volume of rainwater consumed during the production process. This is particularly relevant for agriculture and

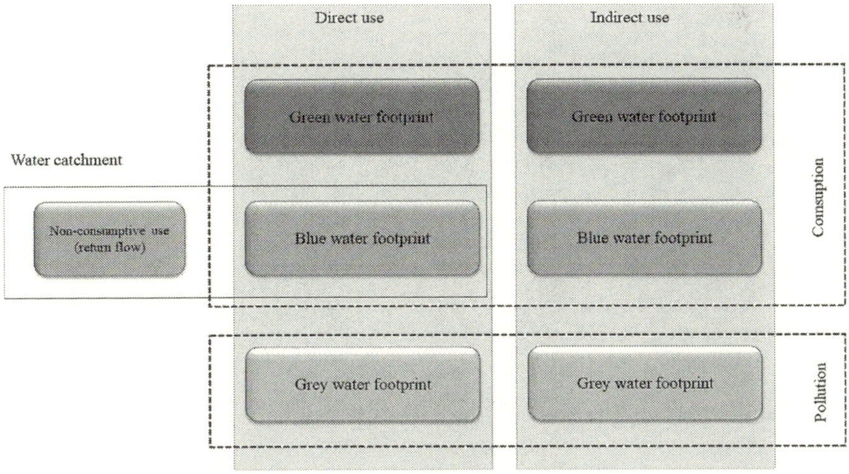

FIG. 3.2 Criteria and divisions of the water footprint.
This figure illustrates the division of direct and indirect water footprint of the processes.

forestry, where it refers to total rainwater (evapotranspiration) plus water incorporated into the crop. Therefore, the green water footprint in one step of the process is quantified according to Eq. (3.24):

$$WF_{green} = GW_e + GW_i \tag{3.24}$$

where WF_{green} is the green water footprint, GW_e is green water evaporation (m^3), and GW_i is green water incorporation (m^3), total expressed in m^3.

It is important to note that the main distinction between the blue water footprint and the green water footprint consists of the hydrological, environmental, and social impacts as well as the costs of surface and groundwater use for production since they differ notoriously from the impacts and costs of rainfall.

Concurrently, the gray water footprint represents a process step which is an indicator of the degree of freshwater pollution that can be associated with this stage of the process. This parameter spread as the proportion was taken that the amount of pollution could be expressed in terms of the volume of water needed to dilute the pollutants so that they would be harmless to the water bodies. Thus, the quantification of the gray water footprint is expressed according to Eq. (3.25):

$$WF_{gray} = \frac{L}{C_{max} - C_{nat}} \tag{3.25}$$

where WF_{gray} is the gray water footprint, L is the pollutant load (kg), C_{max} is the ambient water quality standard for this pollutant (kg/m^3), and C_{nat} is the natural concentration of this pollutant in the water receiver (kg/m^3), expressed in m^3.

Therefore, a brief water footprint category outline can be seen in Table 3.12, and a calculation exemplification is inserted in Box 3.7.

TABLE 3.12 Water footprint category outline.

Impact category	Water footprint (depletion of abiotic resources)
Definition	Decrease in the availability of nonbiological resources (non- and renewable) as a result of their unsustainable use
Impact indicator	Decrease of resources
Considerations	Distinctions between renewable and nonrenewable resources
Damage categories	Resources and possible ecosystem collapse
Unit	m^3

BOX 3.7 Calculation application example for water footprint category.

The water footprint (WF) is generally applied to quantify the demand for water consumed in a given process. However, in some cases, it can also be applied to identify a specific point in the industry, aimed at reducing consumption. In this way, WF analysis helps to identify intensive operations in the consumption of different types of water: blue, green, and gray.

In a biotechnological industrial process based on microalgae, the microalgal cultivation stage is identified as the most water-intensive, and this volume is directly associated with the evaporation rates and the volume incorporated into the biomass. Therefore, the quantification of the total water footprint does not include the gray water footprint, since this is related to the amount of water used to dilute a polluting load that reaches a body of water, based on its natural conditions.

Thus, at the level of exemplification, Severo et al. (2020) evaluated the total water footprint of an integrated microalgal cultivation system. The work volume of the reactor was considered 1 m^3, operating for 330 days a year. Therefore, the total water footprint can be given by adding a blue and green footprint.

In this way, the blue footprint of the process is approximately 47 m^3/year (1 m^3 for every 7 days of cultivation batch). Considering that on average 20% of the cultivated water is incorporated into the biomass, about 9.4 m^3/year of green water is used in the process. Evaporation resulted in a water loss of 0.094 m^3/year, which is attributed to the rate of aeration. Therefore, the total water footprint was estimated at about 56.4 m^3/year.

$$WF_{blue} = BW_e + BW_i + L_{rf}$$

$$WF_{blue} = 1\,\mathrm{m}^3 \times \left(\frac{330\,\mathrm{d}}{7\,\mathrm{d}}\right)$$

$$WF_{blue} = 47\,\mathrm{m}^3$$

$$WF_{green} = GW_e + GW_i WF_{green} = 0.094\,\mathrm{m}^3 + 9.4\,\mathrm{m}^3$$

$$WF_{green} = 9.49\,\mathrm{m}^3$$

$$WF = \sum WF_{blue} + WF_{green} WF = \sum 47\,\mathrm{m}^3 + 9.49\,\mathrm{m}^3\ WF = 56.4\,\mathrm{m}^3$$

On the other hand, a wastewater treatment process may be suitable for quantifying the gray footprint. For example, the final disposal of effluent from slaughter and processing of poultry and swine activity has a nitrogen concentration of 128 mg/L N. However, the legislation suggests that the maximum allowable acceptable value is 2.9 mg/L N, while the natural concentration is approximately 0.4 mg/L N. Therefore, when applying the values in Equation:

$$WF_{gray} = \frac{L}{C_{max} - C_{nat}}$$

$$WF_{gray} = \frac{128.0}{2.9 - 0.4}$$

$$WF_{gray} = 51.2$$

Therefore, the gray hydraulic footprint is 51.2 L. This means that for each 1 L of effluent, it will be necessary to dilute it in 51.2 L, so that the effluent can be disposed of in water resources.

Land use

As the human population is steadily increasing, productive land is becoming an even more limited resource. As a result, land use and land use change have several environmental impacts. In principle, the focus of environmental impacts associated with land use was centered on greenhouse gas emissions, changes in biogeochemical cycles, and carbon storage. More recently, concerns about soil quality, net soil productivity, and biodiversity loss have received high progression in terms of environmental relevance. Nevertheless, research indicates that changes in land use and land cover also impact the quality and availability of water (Fig. 3.3).

Generally, by applying the land use assessment tool to the life cycle assessment, four protection areas are proposed, among them: the protection of human health, the natural environment, natural resources, and the environment modified by man. For the natural environment area, impacts were proposed for an assessment on life support functions and ecosystem services. However, impacts on the naturalness of the areas include the time needed for regeneration between their different types of use and their impacts on biodiversity.

In the current state, although the different terminologies and methodologies for assessing the impact of land use and land use change are still under development, two broad divisions are classified: direct land use changes (dLUC) and indirect changes in the use of land ground (iLUC). Direct land use change is defined primarily as changes in human use or land management within the boundaries of the product system being evaluated, while indirect land use change (iLUC) as changes in the use or management of the land that is a

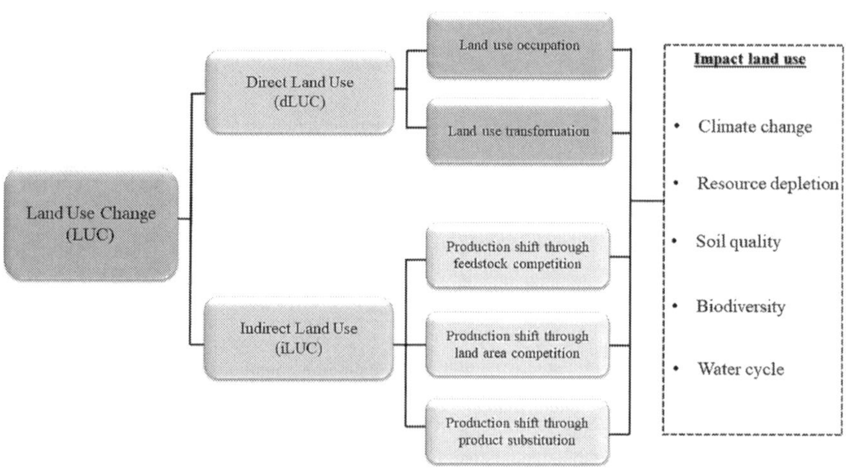

FIG. 3.3 Environmental impacts associated with direct and indirect land use.
This figure illustrates the division of environmental impacts associated with direct and indirect soil emissions.

consequence of the direct change in land use, but which occurs outside the product system being evaluated.

In addition, to the extent that indirect quantification of land use changes presented inconsistencies in the standardization of methods, there is a consensus that in practice its application is still rare and limited to a single environmental impact. In this way, the main land use change assessments are straightforward. Thus, the profiles evaluated consist of two categories: land use transformation and land use occupation. In sum, the change in land use is man-made transformation (e.g., from forests to agricultural production). However, their consequences can lead to high damage such as a decline in biodiversity. From another perspective, it may have a favorable influence, as there may be transformations of built-up areas in gardens or secondary forests (Perminova, Sirina, Laratte, Baranovskaya, & Rikhvanov, 2016).

On the other hand, land occupation is defined by the continuous use of some area for a given time period for identified land use quality. In addition, as well as land use change, the land occupation can also have positive or negative impacts. In this context, all human impacts on soil, water, plant and animal life, when caused by land use activities, are therefore declared as land use impacts. Nonetheless, land use impacts not only assume the ecological consequences, but also form part of a global research agenda, as sustainable land management is also a political, economic, and social issue.

Besides the set of details that ground the land use change impact category, the land occupation comprises the duration in which the area of transformed land returns to its primary natural state. This measure is simplified in terms of the time required to recover the total productivity of the land, in land area (Fthenakis & Kim, 2009). However, the restoration time is highly variable depending on the suspended ecosystem (Table 3.13).

Therefore, to quantify the impacts of land transformation and occupation, one of the recommended methodologies is in accordance with UNEP/SETAC (Benoît Norris et al., 2013) guidelines on assessing the impact of land use on biodiversity (Lathuillière, Miranda, Bulle, Couto, & Johnson, 2017). This, according to Eqs. (3.26) and (3.27) (De Baan, Mutel, Curran, Hellweg, & Koellner, 2013):

$$LU_{trans} = CF_{trans} \times A \tag{3.26}$$

$$LU_{occ} = CF_{tocc} \times A \tag{3.27}$$

where LU_{trans} and LU_{occ} are, respectively, the land transformation and occupation impacts, A is the area transformed or occupied (m^2), A_{tocc} is the occupation time (year/m^2), CF_{trans} is the characterization factor for land transformation (m^2 eq), and CF_{occ} is the characterization factor defined for land occupation (m^2 eq) and shown in Eqs. (3.28) and (3.29).

TABLE 3.13 The estimate of time of restoration for different ecosystems.

Ecosystems	Time (years)
Arable land, pioneer vegetation	<5
Species-poor meadows, mature pioneer vegetation	5–25
Species-poor immature shrubs and hedgerows, oligotrophic vegetation, species-rich marshland, meadows, dry meadows, and heathland	25–50
Species-rich forests, shrubs, and hedgerows	50–200
Immature peat bogs, old dry meadows, and heathland	200–1000
Mature peat bogs, old growth forests	1000–10,000

Based on Fthenakis and Kim (2009).

$$CF_{trans} = \frac{1}{2}(Q_{NV} - Q_{LU}) \times t_{regen} \tag{3.28}$$

$$CF_{occ} = (Q_{NV} - Q_{LU}) \tag{3.29}$$

where CF_{transf} is the characterization factor for land transformation (PDF/m^2), CF_{occ} is the characterization factor for land occupation (PDF/m^2), Q is the ecosystem quality of natural vegetation (Q_{NV}) or new land use (Q_{LU}) (PDF), and t_{regen} is the regeneration time needed for the landscape to relax to natural vegetation (NV) (year/m^2), expressed in m^2.

Therefore, a brief land use category outline can be seen in Table 3.14, and a calculation exemplification is inserted in Box 3.8.

TABLE 3.14 Land use category outline.

Impact category	Land use
Definition	Impact on the land due to agriculture, anthropogenic settlement, and resource extractions
Impact indicator	Species loss, soil loss, amount of organic dry matter content, etc.
Considerations	Analysis of the land area to be altered and observations of biodiversity that could be damaged
Damage categories	Resources or nonrenewable depletion
Unit	Probability density function (PDF)/m^2 or m^2

BOX 3.8 Calculation application example for land use category.

The impact potential associated with the land use category is usually determined to quantify the potential for transformation or occupation that given land use will cause.

In terms of exemplification, the increase in soy production coincided with an evolution in the use of land, water, and fertilizers in the production locations. As a consequence, this excessive demand for natural resources results in the loss of biodiversity and a potential ecosystem imbalance.

The following example suggests the evaluation of the potential of regional species loss, taking into account birds as species in potential extinction, as a result of agricultural activity present in South America.

Thus, it is known that the soybean crops already established represent a territorial occupation of about 22 million hectares, only in Brazilian territory. However, the latest studies estimate that the apt expansion in South America will reach numbers exceeding 100 million hectares. However, it is important to highlight that, for this specific region, the factors for characterization of land transformation and occupation are 4.4×10^9 and 0.47 PDF (for other factors, see DOI: https://doi.org/10.1021/es400592q). Thus, to assess the potential impact associated with the land transformation, it is given through the following calculations:

$$LU_{trans} = CF_{trans} \times A$$

$$LU_{trans} = 4.4 \times 10^9 \text{ PDF} \times 22{,}000{,}000 \text{ ha}$$

$$LU_{trans} = 9.68 \times 10^{16} \text{ PDF}$$

Likewise, the land occupation potential is determined using the following equation:

$$LU_{occ} = CF_{occ} \times A$$

$$LU_{occ} = 0.47 \text{ PDF} \times 22{,}000{,}000 \text{ ha}$$

$$LU_{occ} = 1.03 \times 10^7 \text{ PDF}$$

Therefore, it is estimated that the potential for regional species loss for birds in agriculture is 9.68×10^{16} and 1.03×10^7 PDF, for transformation and occupation, respectively. However, note that the characterization factors were used according to the database provided by De Baan et al. (2013) (see DOI: https://doi.org/10.1021/es400592q).

Global warming potential

Global warming addresses the effect of increasing temperature in the lower atmosphere. The atmosphere is normally heated by radiation received from the sun. A part of the radiation is usually reflected by the Earth's surface, but the carbon content in carbon dioxide (CO_2) and other greenhouse gases reflects or absorbs the radiation resulting in the greenhouse effect, i.e., a rise in temperature in the lower atmosphere to an above-normal level. Among its consequences, highlight is the increase in the temperature level, resulting in the melting of the polar caps, and therefore, increasing the high levels of the sea. In addition, raising the temperature level can also result in regional climate change. Therefore, knowing and quantifying the real impact of emissions is fundamentally important for monitoring the sustainability and remediation of global warming potential (Jolliet et al., 2018).

Since the last decade, life cycle assessment has obtained more space as the most modern and complete data generator used to quantify global warming potential processes, products, or activities. Usually, global warming potential is quantified through the sum of emissions of greenhouse gases' (GHG emissions') impact. It is important to reinforce that substances contributing to global warming are defined as gaseous substances which have infrared absorption capacity or are degraded in CO_2, in addition, have a sufficient lifetime in the atmosphere to contribute significantly to global warming.

The characterization factors of these substances are presented in the literature in common methods, proposed for the temporal evaluations in the environmental evaluation with different equivalences. In terms of global warming potential (GWP) and most commonly CO_2-eq, the factors of characterization for environmental issues are published annually in reports such as the Intergovernmental Panel on Climate Change (IPCC) and models used in the calculations are developed. In addition, recently new metrics were entered as global temperature potential (GTP) and some applications were made for time-adjusted warming potentials (TAWPs) (Table 3.15).

Currently, the following principles are recommended for the selection of short- to long-term assessment criteria for global warming potential categories: (i) GWP100 for short-term impact category when associated with temperature change and (ii) GTP100 for long-term impact category referents to temperature increase. In addition, it is strongly advised to apply the combined use of the two categories of impact to a wider range of data discussion. However, the use of these subcategories is presented as a novel element in assessing global warming potential, and this exercise is traditionally based on the use of a single indicator of climate change, usually the GWP100.

Finally, the most comprehensive and usual model for quantifying global warming potential is described by Laratte, Guillaume, Kim, and Birregah (2014) in Eq. (3.30):

$$GWP = \sum_i GWP_i \times m_i \tag{3.30}$$

where GWP is the global warming potential, i is the time horizon (year), GWP_i is the equivalence factor for the substance i (kg eq), and m_i is the emission of the substance i (kg), expressed in kg CO_2-eq.

Therefore, a brief global warming potential category outline can be seen in Table 3.16, and a calculation exemplification is inserted in Box 3.9.

TABLE 3.15 Characterization factors for selected greenhouse gases, in shorter-term (GWP100) and long-term (GTP100) climate change impacts.

Greenhouse gases	Chemical formula	Lifetime (years)	Shorter-term climate change (GWP100) (kg CO_2-eq.(shorter)/kg)	Long-term climate change (GTP100) (kg CO_2-eq.(long)/kg)
Carbon dioxide	CO_2	Indefinite	1	1
Methane biogenic	Biogenic CH_4	12.4	34	11
Methane fossil	Fossil CH_4	–	36	13
Nitrous oxide	N_2O	121	298	297
HCF-134a	CH_2FCF_3	13.4	1550	530
CFC-11	CCl_3F	45	5350	3490
PFC-14	CF_4	50,000	7350	9560
Sulfur hexafluoride	SF_6	3200	26,087	33,631

Based on Myhre et al. (2013).

TABLE 3.16 Global warming potential category outline.

Impact category	Global warming potential
Definition	Change of global temperature caused by greenhouse gases
Impact indicator	Disturbances in global temperature and climatic phenomenon
Considerations	Greenhouse gases and their global warming potential (GWP)
Damage categories	Change climate
Unit	kg CO_2 equivalent

BOX 3.9 Calculation application example for global warming potential category.

Generally, industrial processes contribute to greenhouse gas emissions in two ways, indirectly through the generation of electricity used to power the systems and directly through exhaust gases, as is the case with combustion systems. However, in some processes, direct emissions can be attributed to some system leaks. A very common example is cooling systems.

In terms of exemplification, vapor compression refrigeration systems use working fluids (refrigerants) that circulate through a closed cycle, changing the phases when passing through heat exchangers. However, these fluids contribute significantly to the destruction of the ozone layer and produce harmful effects on the global environment, especially on the potential for global warming.

Thus, considering a refrigeration system for commercial cold rooms, with an energy demand of approximately 40kW, with a cooling load of 13.8kg of CFC-11, and a leakage rate of 8%, during 24 hours of operation, what would be the heating potential of the total global (direct and indirect) environment of the system?

The first step is to identify the CFC-11 characterization factor (see Table 3.15) and electricity (according to the source of electricity and location of the plant). In this case, we consider electricity from coal (about 0.545 kg CO_2 eq).

$$GWP_{indirect+direct} = \sum_i GWP_i \times m_i$$

$$GWP_{indirect} = \sum_{100\,yr} 0.545\,kg\,CO_2\,eq \times 40\,kW \times 24\,h$$

$$GWP_{indirect} = 523.2\,kg\,CO_2\,eq$$

$$GWP_{direct} = \sum_{100\,yr} 5350\,kg\,CO_2\,eq \times [13.8\,kg\,(8\%)]$$

$$GWP_{direct} = 5906.4\,kg\,CO_2\,eq$$

$$GWP_{indirect+direct} = 523.2\,kg\,CO_2\,eq + 5906.4\,kg\,CO_2\,eq$$

$$GWP_{indirect+direct} = 6431.4\,kg\,CO_2\,eq$$

Therefore, the indirect CO_2 emissions for the system were 523.2 kg CO_2 eq. Additionally, direct CO_2 emissions were 5906.4 kg CO_2 eq, totaling a global warming potential for the process of 6431.4 kg CO_2 eq.

Human toxicity potential

Human toxicity potential (HTP) is a quantitative toxic equivalency potential (TEP) that has been introduced in life cycle assessment for expressing the potential damage of a chemical unit released into the environment. This category is recognized for it includes inherent toxicity and generic source-to-dose ratios for pollutant emissions.

Ideally, all emissions of substances potentially affecting human health should be quantified. However, the plurality of known substances (>100.000) and the even larger number of emission sources logically make that approach unfeasible. Given this aspect, the main substances are grouped according to the respective areas of impact (Table 3.17).

A wide variety of methodologies have been developed and published to quantify human toxicological potentials. Among the developed methods are IMPACT 2002, USES-LCA, Eco-Indicator 99, and CalTOX. These methods adopt environmental multimedia models with multiple paths to account for environmental fate and exposure processes. However, characterization methods, such as the EDIP (Hauschild & Wenzel, 1998), consider fate and exposure based on the key chemical properties.

Still, comparative studies evaluating the different methodologies regarding the quality seem to have little or no difference. However, maybe the main difference between the methodologies used are the units to be expressed. For example, it is possible to directly compare the relative importance of human toxicity potential when estimated in categories of human health damage using a single indicator, such as disability-adjusted life years (DALYs), i.e., a DALY represents 1 year of the emission total (Hofstetter & Hammitt, 2002), whereas, for other methodologies of quantification of the impact categories, 1,4-dichlorobenzene is taken as the reference substance. Thus, to use a methodology simple enough to be used worldwide for a large number of substances, the impact potentials for human toxicity can be calculated according to Eq. (3.31):

$$THP_{hum,m} = \sum\nolimits_{i} EQF \times m_i \tag{3.31}$$

where $THP_{hum,m}$ is the effect potential (human toxicology) caused by exposure via the medium m (water, soil, air), EQF is the equivalency factor for the substance i caused by exposure via the medium m (kg eq), and m_i is the emission of substance i (kg), expressed in kg 1,4-DB-eq.

Therefore, a brief human toxicity potential category outline can be seen in Table 3.18, and a calculation exemplification is inserted in Box 3.10.

Categories of damage or areas of protection (endpoints)

The categories of damage or categories of protection areas are considered the last and optional stage of the life cycle assessment (Pennington et al., 2004). They are characterized as a result of environmental indexes aggregated by

TABLE 3.17 Potentially toxic substances and their quantification bottlenecks.

	Substances	Bottleneck
Air	- Persistent organic pollutants (POPs); - High production volume (HPV).	- Quantification of emissions of chemicals other than heavy metals, POPs, and other traditional atmospheric pollutants (such as NO_x and SO_2) is limited.
Volatile organic compounds (VOCs)	- VOCs from point sources (specific substances/ solvents); - Combustion of VOCs (alkanes, methane, olefins, ethene and monocyclic aromatic, aldehydes and formaldehyde); - Traces of polycyclic aromatic hydrocarbons (PAHs) (among others benzo(α)pyrene and naphthalene).	- Difficulty of assigning a specific toxicity to VOCs as a group as it covers a wide range of substances.
Particles	- Total suspended particulates (TSPs); - Particles with a diameter below 10 μm (PM10).	- The measurement data for this parameter are still very incomplete.
Water	- Deposition of heavy metals and POPs and other chemicals from air; - Emissions from sewage treatment plants; detergents, metals, and others.	- Mistaking of quantification referents with the ecotoxicity potential (EP).
Soil	- Deposition of heavy metals, POPs, and other chemicals emitted to air; - Sludge from sewage treatment plants.	

TABLE 3.18 Human toxicity potential category outline.

Impact category	Human toxicity
Definition	Toxic effects of chemicals on humans
Impact indicator	Cancer, respiratory diseases, other noncarcinogenic effects and effects on ionizing radiation
Considerations	Toxicological responses of humans and nature of the chemicals in the human body
Damage categories	Human health
Unit	kg 1,4-DB equivalent or DALY (Disability-adjusted life year)

BOX 3.10 Calculation application example for human toxicity potential category.

In industrial manufacturing processes, impacts on human health generally refer to the influence of residues, effluents, emissions, and industrial byproducts on human beings. As a consequence, these indicators indicate the degree of risk that these emissions represent for the well-being and continuity of human life.

Evidently, the potential for human toxicity of residues from productive operations differs according to the nature of the operations necessary for the production of its products. Among the industries that present residues and bioproducts potentially impacting in relation to human toxicity are the sugar and alcohol industry, which, through the processing of sugarcane, is responsible for the production of ethanol, sugar, and energy, beyond other marketable bioproducts.

Thus, at the level of exemplification, a sugar-energy industrial plant releases waste/bioproducts such as vinasse, filter cake, bagasse, gas flaring, tank washing effluent, among others, which contain potentially toxic elements such as cadmium (Cd), nickel (Ni), zinc (Zn), iron (Fe), and lead (Pb). These elements can have a negative impact on human health if they are not disposed of properly.

In the case of filter cake, it is estimated that on average the plants produce 32 kg to 42 kg/ton of processed cane. In addition, among the toxic elements, cadmium represents about 2.55 mg/kg of cake, totaling about 81.6 mg of cadmium per ton of processed sugar cane.

However, the human body accumulates cadmium and at the age of 50, the man may have a charge of 20–30 mg, concentrating on the kidneys and artery walls. More serious effects are due to this high concentration of cadmium: the destruction of testicular tissue and red blood cells. Thus, for every 1 ton of raw material processed, the cadmium values spent on the waste are twice as high as the accumulative limits in the human body.

When consulting the individual potential values specific to cadmium, it is estimated that the characterization factor of this compound is $5.1 \times 10^{+2}$ kg 1,4-DB-eq.

Continued

BOX 3.10 Calculation application example for human toxicity potential category—cont'd

Thus, considering 1 kg of processed sugar cane, only the filter cake can have a human toxicity potential of 0.04 kg 1,4-DB-eq, as follows:

$$THP_{hum,m} = \sum_i EQF \times m_i$$

$$THP_{hum,m} = \sum_i 5.1 \times 10^{+2}\,\text{kg}\,1,4-\text{DB}-\text{eq} \times 8.16 \times 10^{-5}\,\text{kg}$$

$$THP_{hum,m} = 0.04\,\text{kg}\,1,4-\text{DB}-\text{eq}$$

impact categories (midpoints). This calculation is performed through weighting factors for the impact categories evaluated (see Chapter 4). Previously, it was established groups such as the use of resources, consequences for human health and ecological. Fig. 3.4 illustrates these groups under the context of endpoint indicators. Thus, damage categories such as human health and natural environment (resources) are described in Table 3.19.

Finally, the final interpretation of endpoint results does not require a thorough knowledge of their environmental effects. In addition, the endpoints present greater statistical uncertainties and some assumptions are accumulated along the chain. On the other hand, there is no defined methodology to standardize the use of midpoints or endpoints. However, even if the midpoint approach is the most used category, the harm categories approach is used because it presents results that are easier to describe, but lower than the midpoint approaches (Bare, Hofstetter, Pennington, & De Haes, 2000).

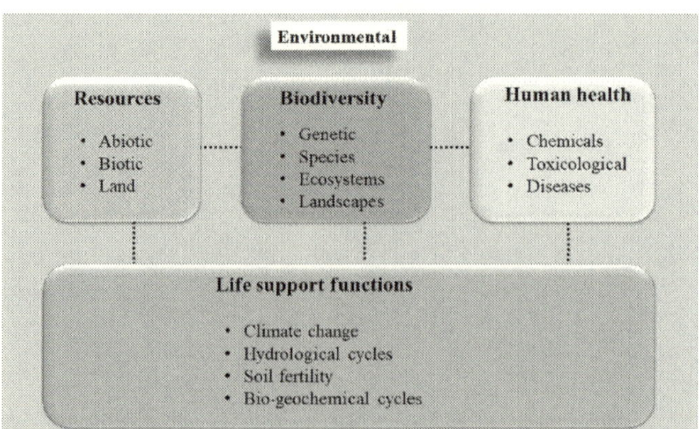

FIG. 3.4 Areas of protection and their impacts on the environment.
This figure illustrates the use of resources, consequences for human health, and ecology under the context of endpoint indicators.

TABLE 3.19 Damage categories and their respective indicators.

Endpoints	Indicators of categories
Ecosystem quality	Several effects (increased fish production capacity, contribution to species extinction); Increased growth of aquatic biomass; Increase of hazard index for ecosystems above critical levels; Increase of potentially disappeared fraction of vegetation species (dealing with terrestrial eutrophication and acidification together);
Resources	Extraction rate and potential extinction risks; Acquisition of near-sustainable alternatives; Future extraction energy; Water availability and scarcity.
Climate change	Increase of radiative forcing (GWPs); Increase of human health effects (heat/cold stress, malaria, dengue fever, disaster, nutrient deficiency); Increase in crop and primary production damage; Increase of several damages—human health and ecosystem effects (heat stress, malaria, starvation, flooding accidents, wood growth, contribution to species extinction).
Human health	Policy-based hazard equivalents using multimedia fate and multipathway exposure modeling with policy-based toxicological thresholds; cancer and respiratory factors, accounting for impacts in terms of DALY

References and recommended reading

Papers of particular interest have been highlighted as:
• of special interest
•• of outstanding interest

Altenstedt, J., & Pleijel, K. (1998). *POCP for individual VOC under European conditions*. Swedish Environmental Research Institute.

Andersson-Sköld, Y., & Holmberg, L. (2000). Photochemical ozone creation potentials (POCP) and replacement of solvents in Europe. *Atmospheric Environment*, *34*(19), 3159–3169.

Arodudu, O., Ibrahim, E., Voinov, A., & van Duren, I. (2014). Exploring bioenergy potentials of built-up areas based on NEG-EROEI indicators. *Ecological Indicators*, *47*, 67–79.

•• Bare, J. C., Hofstetter, P., Pennington, D. W., & De Haes, H. A. U. (2000). Midpoints versus endpoints: The sacrifices and benefits. *The International Journal of Life Cycle Assessment*, *5*(6), 319. This document reported the implications of mid-point versus end-point indicators for uncertainty, transparency, and the ability to subsequently resolve offsets between impact categories using weighting techniques.

Benoît Norris, C., Traverso, M., Valdivia, S., Vickery-Niederman, G., Franze, J., Azuero, L., & Aulisio, D. (2013). *The methodological sheets for sub-categories in social life cycle assessment (S-LCA)*. United Nations Environment Programme (UNEP) and Society for Environmental Toxicology and Chemistry (SETAC). This document presents the history of social life cycle assessment, the project group, and the drafting process of the guidelines. In addition, it shows the main contributions to the development of methodologies and their main characteristics and usability of the evaluation of the social life cycle. Also, they report the depth and breadth of the methodology compared to other environmental tools.

De Baan, L., Mutel, C. L., Curran, M., Hellweg, S., & Koellner, T. (2013). Land use in life cycle assessment: Global characterization factors based on regional and global potential species extinction. *Environmental Science & Technology*, *47*(16), 9281–9290.

Derwent, R. G., Jenkin, M. E., Passant, N. R., & Pilling, M. J. (2007). Reactivity-based strategies for photochemical ozone control in Europe. *Environmental Science & Policy*, *10*, 445–453.

Derwent, R. G., Jenkin, M. E., & Saunders, S. M. (1996). Photochemical ozone creation potentials for a large number of reactive hydrocarbons under European conditions. *Atmospheric Environment*, *30*(2), 181–199.

• EEA. European Environment Agency. (2000). *Environment in the European Union at the turn of the century: Offprint: Greenhouse gases and climate change*. Office for Official Publ. of the European Communities. This document reports the guidelines are supported by a number of technical reports, and which present the scientific discussions and documentation for recommendations in the field of the life cycle assessment.

Farid, M. A. A., Roslan, A. M., Hassan, M. A., Hasan, M. Y., Othman, M. R., & Shirai, Y. (2020). Net energy and techno-economic assessment of biodiesel production from waste cooking oil using a semi-industrial plant: A Malaysia perspective. *Sustainable Energy Technologies and Assessments*, *39*, 100700.

Fenger, J. (2009). Air pollution in the last 50 years—From local to global. *Atmospheric Environment*, *43*(1), 13–22.

Fore, S. R., Porter, P., & Lazarus, W. (2011). Net energy balance of small-scale on-farm biodiesel production from canola and soybean. *Biomass and Bioenergy*, *35*(5), 2234–2244.

Fthenakis, V., & Kim, H. C. (2009). Land use and electricity generation: A life-cycle analysis. *Renewable and Sustainable Energy Reviews*, *13*(6–7), 1465–1474.

García, C. A., Fuentes, A., Hennecke, A., Riegelhaupt, E., Manzini, F., & Masera, O. (2011). Life-cycle greenhouse gas emissions and energy balances of sugarcane ethanol production in Mexico. *Applied Energy*, *88*(6), 2088–2097.

Guinée, J. B., & Lindeijer, E. (Eds.). (2002). *Vol. 7. Handbook on life cycle assessment: Operational guide to the ISO standards* Springer Science & Business Media.

Hall, C. A. (1972). Migration and metabolism in a temperate stream ecosystem. *Ecology*, *53*(4), 585–604.

Hauschild, M., & Wenzel, H. (1998). Stratospheric ozone depletion as a criterion in the environmental assessment of products. In M. Hauschild, & H. Wenzel (Eds.), *Scientific background: Vol. 2. environmental assessment of products*. London: Chapman & Hall.

Hauschild, M. Z., Goedkoop, M., Guinée, J., Heijungs, R., Huijbregts, M., Jolliet, O., & Sala, S. (2013). Identifying best existing practice for characterization modeling in life cycle impact assessment. *The International Journal of Life Cycle Assessment*, *18*(3), 683–697.

Hill, J., Nelson, E., Tilman, D., Polasky, S., & Tiffany, D. (2006). Environmental, economic, and energetic costs and benefits of biodiesel and ethanol biofuels. *Proceedings of the National Academy of Sciences*, *103*(30), 11206–11210.

• Hoekstra, A. Y., & Mekonnen, M. M. (2012). The water footprint of humanity. *Proceedings of the National Academy of Sciences*, *109*(9), 3232–3237. This study quantifies and maps the water footprint of humanity at a high spatial resolution. It reports on consumptive use of rainwater and ground and surface water and volumes of water polluted. Water footprints are estimated per nation from both a production and consumption perspective. The relevance this paper is to illustrate the global dimension of water consumption and pollution by showing that several countries heavily rely on foreign water resources and that many countries have significant impacts on water consumption and pollution elsewhere.

Hofstetter, P., & Hammitt, J. K. (2002). Selecting human health metrics for environmental decision-support tools. *Risk Analysis, 22*(5), 965–983.

ILDC. (2010). International Reference Life Cycle Data System Handbook—General Guide for Life Cycle Assessment—Detailed Guidance. In *EUR 24708 EN* (1st edition). Luxembourg: Publications Office of the European Union.

● ISO. (2014). *14046 environmental management, water footprint—Principles, requirements and guidelines.* Geneva: International Organization for Standardization. This normative specifies principles, requirements, and guidelines related to water footprint assessment of products, processes, and organizations based on life cycle assessment.

Jenkin, M. E., & Hayman, G. D. (1999). Photochemical ozone creation potentials for oxygenated volatile organic compounds: Sensitivity to variations in kinetic and mechanistic parameters. *Atmospheric Environment, 33*(8), 1275–1293.

Jolliet, O., Margni, M., Charles, R., Humbert, S., Payet, J., Rebitzer, G., & Rosenbaum, R. (2003). IMPACT 2002+: A new life cycle impact assessment methodology. *The International Journal of Life Cycle Assessment, 8*(6), 324.

Jolliet, O., Antón, A., Boulay, A. M., Cherubini, F., Fantke, P., Levasseur, A., & Pfister, S. (2018). Global guidance on environmental life cycle impact assessment indicators: Impacts of climate change, fine particulate matter formation, water consumption and land use. *The International Journal of Life Cycle Assessment, 23*, 2189–2207.

Jorquera, O., Kiperstok, A., Sales, E. A., Embirucu, M., & Ghirardi, M. L. (2010). Comparative energy life-cycle analyses of microalgal biomass production in open ponds and photobioreactors. *Bioresource Technology, 101*(4), 1406–1413.

Kamahara, H., Hasanudin, U., Widiyanto, A., Tachibana, R., Atsuta, Y., Goto, N., … Fujie, K. (2010). Improvement potential for net energy balance of biodiesel derived from palm oil: A case study from Indonesian practice. *Biomass and Bioenergy, 34*(12), 1818–1824.

Laratte, B., Guillaume, B., Kim, J., & Birregah, B. (2014). Modeling cumulative effects in life cycle assessment: The case of fertilizer in wheat production contributing to the global warming potential. *Science of the Total Environment, 481*, 588–595.

Lathuillière, M. J., Miranda, E. J., Bulle, C., Couto, E. G., & Johnson, M. S. (2017). Land occupation and transformation impacts of soybean production in Southern Amazonia, Brazil. *Journal of Cleaner Production, 149*, 680–689.

Malça, J., & Freire, F. (2006). Renewability and life-cycle energy efficiency of bioethanol and bio-ethyl tertiary butyl ether (bioETBE): Assessing the implications of allocation. *Energy, 31*(15), 3362–3380.

Mayer, F. D., Brondani, M., Carrillo, M. C. V., Hoffmann, R., & Lora, E. E. S. (2020). Revisiting energy efficiency, renewability, and sustainability indicators in biofuels life cycle: Analysis and standardization proposal. *Journal of Cleaner Production, 252*, 119850.

● Mekonnen, M., & Hoekstra, A. Y. (2011). *National water footprint accounts: The green, blue and grey water footprint of production and consumption* (Value of water research report 50; No. 50). Delft: UNESCO-IHE Institute for Water Education. This study quantifies and maps the water footprints of nations from both a production and consumption perspective and estimates international virtual water flows, and national and global water savings as a result of trade.

Mourad, A. L., & Walter, A. (2011). The energy balance of soybean biodiesel in Brazil: A case study. *Biofuels, Bioproducts and Biorefining, 5*(2), 185–197.

Myhre, et al. (2013). Anthropogenic and natural radiative forcing. In *IPCC (2013) WGI climate change: The scientific basis.* Cambridge, UK, and New York, USA: Cambridge University Press (chapter 8).

●● Pennington, D. W., Potting, J., Finnveden, G., Lindeijer, E., Jolliet, O., Rydberg, T., & Rebitzer, G. (2004). Life cycle assessment Part 2: Current impact assessment practice. *Environment International*, *30*(5), 721–739. This paper highlights how practitioners and researchers from many domains have come together to provide indicators for the different impacts attributable to products in the life cycle impact assessment phase.

●● Perminova, T., Sirina, N., Laratte, B., Baranovskaya, N., & Rikhvanov, L. (2016). Methods for land use impact assessment: A review. *Environmental Impact Assessment Review*, *60*, 64–74. This review paper presents different methods of assessing the impact of land use. Comparison of the methods allowed to identify their specific characteristics and to conclude that the combination of several methods is the best basis for a comprehensive analysis of land use impact assessment.

Pradhan, A., Shrestha, D. S., Van Gerpen, J., & Duffield, J. (2008). The energy balance of soybean oil biodiesel production: A review of past studies. *Transactions of the ASABE*, *51*(1), 185–194.

Reuss, J. O., & Johnson, D. W. (2012). *Vol. 59. Acid deposition and the acidification of soils and waters*. Springer Science & Business Media.

Severo, I. A., Deprá, M. C., Dias, R. R., Barin, J. S., de Menezes, C. R., Wagner, R., … Jacob-Lopes, E. (2020). Bio-combustion of petroleum coke: The process integration with photobioreactors. Part II–Sustainability metrics and bioeconomy. *Chemical Engineering Science*, *213*, 115412.

Shapouri, H., Duffield, J. A., & Wang, M. Q. (2002). The energy balance of corn ethanol: An update. *Agricultural Economic Report No. 814* (pp. 1–19).

Sheehan, J., Camobreco, V., Duffield, J., Graboski, M., Graboski, M., & Shapouri, H. (1998). *Life cycle inventory of biodiesel and petroleum diesel for use in an urban bus (No. NREL/SR-580-24089)*. Golden, CO: National Renewable Energy Lab (NREL).

Sun, C. H., Fu, Q., Liao, Q., Xia, A., Huang, Y., Zhu, X., … Chang, H. X. (2019). Life-cycle assessment of biofuel production from microalgae via various bioenergy conversion systems. *Energy*, *171*, 1033–1045.

Tanaka, P. L., Oldfield, S., Neece, J. D., Mullins, C. B., & Allen, D. T. (2000). Anthropogenic sources of chlorine and ozone formation in urban atmospheres. *Environmental Science & Technology*, *34*(21), 4470–4473.

● Tillman, A. M. (2000). Significance of decision-making for LCA methodology. *Environmental Impact Assessment Review*, *20*(1), 113–123. This document specifies the steps of inventory and impacts analysis of life cycle assessment. In addition, it describes scope activities and presents the development of guidelines and principles for implementing a product life cycle assessment.

Vaal, M. A., & Folkerts, A. J. (1998). Sensitivity of microscale ecotoxicity tests and their suitability to measure toxicity of environmental samples. *RIVM Rapport 607042009*.

Chapter 4

Optional elements of life cycle impact assessment

Once inventory data have been properly characterized against their respective impact categories, the results are strongly encouraged to be submitted to three optional elementary categories of assessment: grouping, normalization, and weighting. These categories have the slant of facilitating the interpretation of the data obtained in the impact categories. In addition, they provide a broader coverage in terms of discussions and comparisons with external data from the scientific literature.

Grouping

The grouping step is defined as a qualitative or semiquantitative process which comprises assigning impact categories to one or more sets. Basically, it aims to classify the categories into two possible approaches: a nominal or ordinal (hierarchical) basis.

The nominal grouping method consists of grouping the categories by similar characteristics, e.g., as to entry and exit requirements, or by global, regional, or local geographic scales. In contrast, the hierarchical grouping method, also known as ranking, comprises assignments of values according to their respective priorities, which may be understood as low-, medium-, or high-intensity impacts. In addition, this classification may provide support that assists in the outcome of the relative importance of different impact categories.

However, regardless of the method to be chosen, it is important to emphasize that in the grouping stage, although starting from the same results of characterized or normalized indicators, it is possible that the final results are different since the person responsible for the analysis can assign values for each category evaluated.

Normalization

Normalization is defined in life cycle assessment, as the stage in which the results of the impact assessment are compared to a given reference. The purpose

Sustainability Metrics and Indicators of Environmental Impact. https://doi.org/10.1016/B978-0-12-823411-2.00008-6

of this step is basically to aid in the interpretation of the results and help in the detection of possible inconsistencies contained in them. In addition, it is able to simplify the communication of results to nonpractitioners, as well as serve as a prerequisite for the other optional phases such as weighting.

Among the principles that encourage the application of normalization, we can highlight an important intrinsic factor of this category, which consists of its conception of the magnitude of an impact. Otherwise, to the extent that the values are assigned to reference situations, the data obtained no longer have only the impact of a particular process or product, but also assume scope of application for comparative assessments in different countries or regions over years. This feature is particularly interesting since, when associated with end-point methods, it enables the identification of relevant emission sources and other resources related to human health, ecosystems, or natural resources.

As seen in Chapter 2, the mathematical model of the normalization consists of variables that show the impact category to be evaluated, the characterized impact of the product being studied, and the reference value in relation to which the results are equated. In addition, in the process of choosing the reference system of normalization, there is a wide influence on the part of the results and the way of conduct when interpreting them. In view of the above, it is important to note that there are two main types of normalization, classified as internal or external.

Essentially, the internal normalization is particularized as a tool to compare the environmental performance of different alternative scenarios. In this way, the same impact results will be compared with a reference from the same study. This action can be conducted in various ways, such as division by baseline or maximum division estimates. In addition, by denotation, the internal normalization becomes strictly applicable to comparative scenarios, since it can generate incongruent data if a later stage of weighting is performed. Therefore, the recommendation of the United Nations Program for the Environment and the Society of Environmental Chemistry and Toxicology is currently used as of external normalization.

On the other hand, in external normalization, the reference standard is usually a geographical area that can be global, national, regional, or local. Another aspect diverging from internal normalization is the detailing of a specific period, usually estimated at 1 year. In addition, this category presents two alternatives to follow the external normalization: production-based approach and consumption-based approach. The production-based approach derives the factors of normalization of direct environmental flows linked to human activities, verified within an established area. In contrast, the consumption-based approach encompasses both direct and indirect environmental flows linked to final consumption. However, if considered global scale, both approaches generate equivalent normalization factors, which reflect the same environmental impact in the world. Therefore, the difficulty of data collection and the extrapolation requirements of global inventories have prevented the development and normalization of the scales used.

To date, references are predetermined to be consistent with system boundaries. However, when considering the market limits and the broad need for import and export of raw materials, the chosen reference system must inevitably comprise as many possible stages of the life cycle as possible. In this sense, the use of global normalization references is strictly recommended.

Different sets of normalization can produce different results in terms of identifying the most relevant impact categories, stages, and life cycle assessment. In this sense, some comparative studies between different geographic areas are performed and made available for the public with the objective of evaluating the interferences of the area between the references (Table 4.1).

TABLE 4.1 Comparison between the global, European, and Denmark reference values and their respective impact categories.

	Normalization reference			
Impact categories	**Units**	**Global**	**EU-15**	**Denmark**
Global warming	ton CO_2-eq./capita/year	8.7	8.7	8.7
Stratospheric ozone depletion	kg CFC-11-eq./capita/year	0.103	0.103	0.103
Photochemical ozone formation	kg C_2H_4-eq./capita/year	22	25	20
Acidification	kg SO_2-eq./capita/year	59	74	101
Nutrient enrichment	kg NO_3-eq./capita/year	95	119	260
Human toxicity (via air)	m^3 air/capita/year	2.45×10^9	3.06×10^9	2.09×10^9
Human toxicity (via water)	m^3 water/capita/year	4.18×10^4	5.22×10^4	1.79×10^5
Human toxicity (via soil)	m^3 soil/capita/year	1.02×10^2	1.27×10^2	1.57×10^2
Ecotoxicity, water, acute	m^3 water/capita/year	2.33×10^4	2.91×10^4	7.91×10^5
Ecotoxicity, water, chronic	m^3 water/capita/year	2.82×10^5	3.52×10^5	7.40×10^4
Ecotoxicity, soil, chronic	m^3 soil/capita/year	7.71×10^5	9.64×10^5	6.56×10^5

Based on Stranddorf, Hoffmann, and Schmidt (2004).

Weighting

Weighting is the last optional stage in the life cycle impact assessment. This step is the most normative of the method, and perhaps the most debated. The weighting is the multiplication of the results derived from the standardization of each impact category by a weighting factor that expresses the proportional relevance of a given category (see Chapter 2). However, the weighting factor for a specific impact category is considered key variables in obtaining the data (Eq. 4.1).

$$WgF_k = \frac{AERY_k}{TETY_k} \tag{4.1}$$

where WgF_k is the weighting factor for a specific impact category, *AERY* is actual emission of category in reference year, *TETY* is target emission for category in target year, and k is the weighting indicative.

Additionally, in this step, there is total standardization of units and they can be incorporated for a single score. Faced with this, we can say that ponder is to apply a value judgment to its results. The justification for using the weighting step lies in the ease of presenting the data in a single score. In short, this alternative allows a simplified comparison of the environmental impacts of different processes or products. Moreover, it supports decision-making as it facilitates observation of results.

At the level of technical reports, weighting can be very advantageous. If we analyze as a whole, the life cycle assessment has many steps and numerous details to describe. Thus, with the weighting, the ease of understanding the results in a single score is undoubtedly greater than elucidating distinct scores by processes or products.

In view of the above, weighting methods can be classified in different ways. These methods usually cover one or more sets of predefined factors, and each methodology has its own point of view to determine them. Thus, although they have their peculiarities, four categories can be determined: the distance to the objective of the policy, distance from the scientific target, monetization, and panel weighting.

Basically, the distance category for the purpose of the policy is based on the determination of the weighting factors based on the political goals, that is, on the objectification of reaching a value in a certain period of time. In general, this weighting method is applied to environmental policies geared to climate change. By way exemplification, if the goal for climate change is to reduce global carbon dioxide emissions by 40%, and the goal of acidification is to reduce by around 15%, the carbon footprint will have a greater weighting factor. Still, it is important to note the point that with this approach, policy goals may not reflect the seriousness of the problem, since the policy can be influenced by costs and other political considerations.

In contrast, the distance to the scientific target consists of the same approach as the distance to the political objective; however, they use scientific targets as a reference. For example, it is established through a consensus that there is a maximum safe level of a given substance in the atmosphere. Thus, these levels are used as a reference point. Therefore, the further the study situation, the higher the weighting level. However, we must take into consideration that not every category of impact presents updated reference points or available scientific data, thus providing an inconsistency of results.

The panel weighting is known as the simplest category of determining the weighting factors. In short, the method is based on the opinion of a large group of people, who should judge which category of impact they consider most important. However, this category has many disadvantages compared with the others, since the results can be naturally distorted by the individual peculiarities of each panelist. This is directly related to the different levels of knowledge, as well as the individual experiences with their respective environmental visions.

Additionally, the monetization category has the characteristic of highlighting the environmental impacts in monetary values. This amount can be accounted against the costs related to the prevention or repair of environmental damages. In this way, a certain environmental impact may require a greater or lesser monetary demand. On the other hand, another possibility of monetization is to quantify the value that individuals are willing to pay to avoid a specific impact.

In this way, given the different possibilities of weighting to date, no concrete way has been established. In this sense, no set of classifications or weighting factors were determined as appropriate. Thus, according to the standard, life cycle assessment is encouraged not to apply this optional element in studies that support comparative statements propagated to the public.

Nevertheless, once we choose the weighting, the data of the main European countries, as well as global references, are made available to the public through technical reports to ensure greater standardization of data (Table 4.2).

Finally, although they are optional, grouping, normalization, and weighting are considered key sustainability indexes for practically quantifying sustainable development. However, we can say that both normalization and weighting still show results at arbitrary levels. Moreover, these indices, despite the objectivity of the methodologies used, are still subject to subjectivity and, consequently, present critical premises complex against standardization. Thus, international efforts are driven to develop sustainability indices to cover environmental, economic, and social aspects to extinguish tendentious and nonrobust environmental policy indicators. Therefore, this would ensure the standardization of the specific indicators involved in the environmental assessment processes, encouraging environmental analysts to insert these indicators into their life cycle assessment.

TABLE 4.2 Reference values of weighting for impact categories in European, Denmark, and global scores.

	Weighting reference			
Impact categories	**Units**	**Global**	**EU-15**	**Denmark**
Global warming	ton CO_2-eq./capita/year	1.12	1.05	1.11
Stratospheric ozone depletion	kg CFC-11-eq./capita/year	4.43	2.46	∞
Photochemical ozone formation	kg C_2H_4- eq./capita/year	1	1.33	1.26
Acidification	kg SO_2-eq./capita/year	n.c.[a]	1.27	1.34
Nutrient enrichment	kg NO_3-eq./capita/year	n.c.[a]	1.22	1.31
Human toxicity (via air)	m^3 air/capita/year	n.c.[a]	1.06	1.11
Human toxicity (via water)	m^3 water/capita/year	n.c.[a]	1.3	1.02
Human toxicity (via soil)	m^3 soil/capita/year	n.c.[a]	1.23	1.02
Ecotoxicity, water, acute	m^3 water/capita/year	n.c.[a]	1.11	1.73
Ecotoxicity, water, chronic	m^3 water/capita/year	n.c.[a]	1.18	1.67
Ecotoxicity, soil, chronic	m^3 soil/capita/year	n.c.[a]	1	1.56

[a]*Weighting factors have not been established worldwide; the European weighting factors are recommended for impact potentials located outside Europe or at unknown locality.*
Based on Stranddorf et al. (2004).

References and recommended reading

Stranddorf, H., Hoffmann, L., & Schmidt, A. (2004). Impact categories, normalisation and weighting in LCA. In *Environmental news no. 77*. Copenhagen: The Danish Ministry of the Environment. Environmental Protection Agency.

Chapter 5

Assistant's tools toward life cycle assessment

Defining sustainable industrial processes and agricultural systems is not simple. The main issues involved in these processes involve energy, feedstocks, mobility, planning, water, and wastes; and its improvements are promoted through different strategies. Thus, the viability of each strategy in environmental, economic, and social terms is fundamental to ensure sustainable global performance.

Given this scenario, several efforts are being applied in research to elucidate potential strategies of quantitative methods, which aim to identify, fill, and improve the gaps associated with sustainable and social issues of industrial processes. In this way, among the main strategies development associates with life cycle assessment are life cycle thinking, life cycle costing, social life cycle assessment, and life cycle sustainability assessment.

Life cycle thinking

Life cycle thinking (LCT) is defined as the way of thinking that includes the economic, environmental, and social consequences of a product or process throughout its life. More precisely, it consists of a theoretical approach that investigates improvements and reductions in the impacts of services and products at all stages of processing, including extraction, conversion, transformation, distribution, use, and final destination (Fig. 5.1).

The advantages of this evaluation approach are based on the ease of avoiding creating new problems when solving others. It also refrains from the so-called displacement of loads, that is, the displacement of impacts from one stage of the life cycle to another.

Additionally, the life cycle of thinking makes it possible to go beyond the traditional focus on the production site and manufacturing processes to include the environmental, social, and economic impacts of a product throughout its life cycle. In this way, this approach can be applied at different scales, from single products to more complex systems. However, the introduction of a function-oriented evaluation paradigm and the consideration of upstream and downstream impacts are widely advocated as analytical benefits of the life cycle

Sustainability Metrics and Indicators of Environmental Impact. https://doi.org/10.1016/B978-0-12-823411-2.00006-2

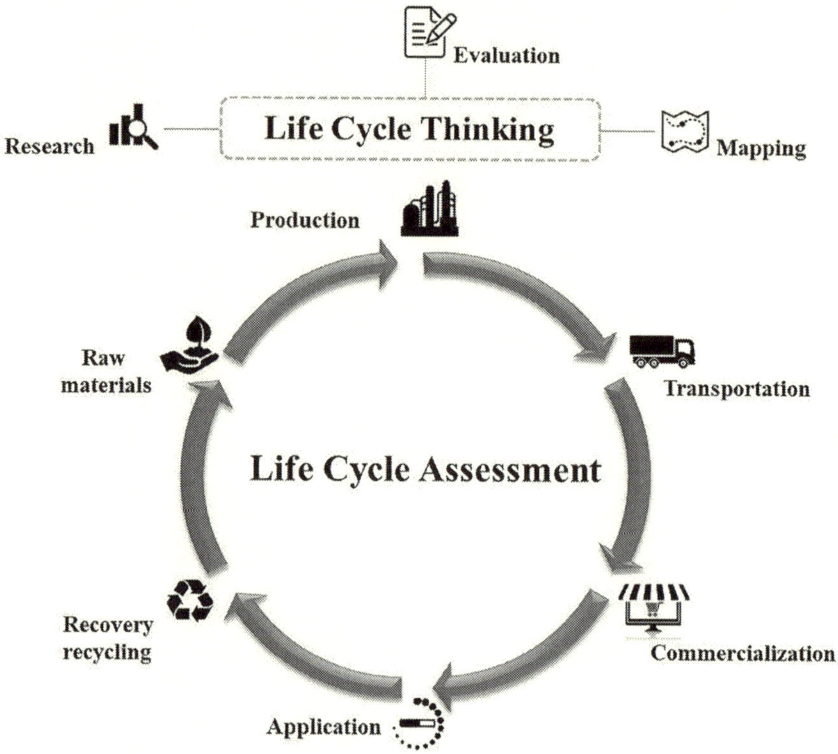

FIG. 5.1 Elements of life cycle thinking.
This figure shows the interaction between the elements that include the life cycle thinking and the life cycle assessment tool.

application of thinking. Therefore, it can be used to complement the hierarchy of the identification needs of environmentally preferable options.

The differential of this approach is presented in the tool used in the strategy to plan the evaluation. In general, there are three main tools for thinking about the life cycle, such as simplified search, mapping, and evaluation.

The research part is based on the exploration and synthesis of the existing data of the life cycle assessment. This research allows the comparison of values of similar systems to the study in question and consequently allows an advance in the total understanding of the analysis, making it broader and deepening the impacts and effects in the entrances and exits of the process.

Just as in the research, the mapping components enable discoveries against the hidden processes included in a production system, as well as compare different functional units to achieve the goal or expected result. The purpose of this step is to help in the broader perspective of the impacts that result in the decisions that are made. In short, we can define this tool as a more thorough reflection of cause and effect. In addition, at the same time, it has the ability to provide perceptions which can change the way the products are processed.

In contrast, simplified assessment tools are designed to assist in rapid life cycle assessment to help inform project decisions and also adjust process impacts. It is important to note that these simplified tools, as well as life cycle mapping, are applied only for the purpose of supporting decisions and should not be used as a full life cycle assessment. Therefore, these tools are only used to increase knowledge and capacity building.

Social life cycle assessment

Alternatively, another approach to life cycle assessment is the social life cycle assessment (S-LCA). On the whole, it is a method that can be used to assess social and sociological aspects of products or processes.

To better understand this goal, it becomes important to define what are a social impact and indicator. As reported in the guidelines and principles for social impact assessment, social impacts are the consequences on human populations of any public or private actions that people live in, work with, play with, relate to, organize themselves to satisfy their needs and, in general, to deal as members of society. On the other hand, indicators are defined as assessments of the social context around the process units.

Given this definition, more specifically, social life cycle impact assessment is the process by which inventory data are aggregated into subcategories and categories to help understand the magnitude of the environmental consequences for the population. In addition, this tool focuses on the elaboration of indicators associated with impact categories to groups such as workers, consumers, the local community, and society associated with value chains (Fig. 5.2).

In conducting a life cycle assessment with a social character, we can consider impact categories as stakeholders, and in this way classify social bottlenecks as subcategories. Thus, in the impact assessment, social and socioeconomic

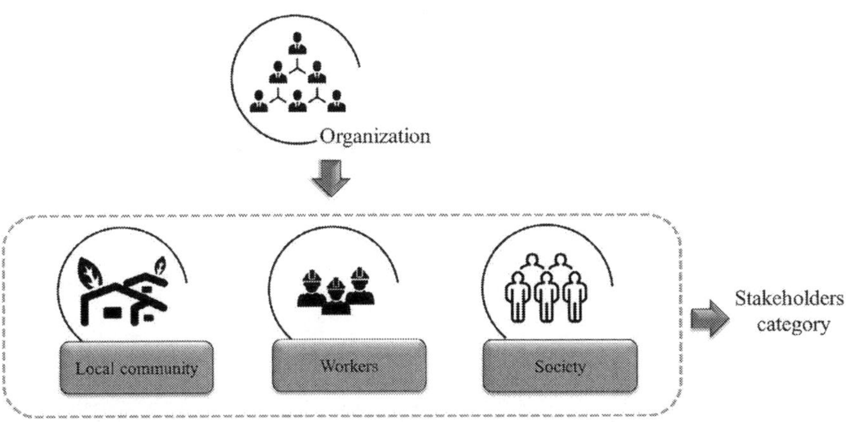

FIG. 5.2 Stakeholder categories related to S-LCA.
This figure addresses the organizational structure of the social life cycle assessment tool, considering the aspects associated with stakeholder categories.

TABLE 5.1 Categories and subcategories guides for stakeholders of the social life cycle assessment.

Stakeholder categories	Subcategories
Workers	- Freedom of association and collective bargaining - Child labor - Fair salary - Working hours - Forced labor - Equal opportunities/discrimination - Health and safety - Social benefits/social security
Consumer	- Health and Safety - Feedback mechanism - Consumer privacy - Transparency - End of life responsibility
Local community	- Access to material resources - Access to immaterial resources - Delocalization and migration - Cultural heritage - Safe and healthy living conditions - Respect of indigenous rights - Community engagement - Local employment - Secure living conditions
Value chain actors	- Fair competition - Promoting social responsibility - Supplier relationships - Respect of intellectual property rights

Based on Andrews and Withey (2012).

subcategories can be inserted according to the categories of impact considered at the beginning of the analysis. Therefore, some reports have established large stakeholder categories and subcategories to be applied as guides for the application of the social life cycle assessment (Table 5.1).

In the current context, social approaches to life cycle assessment are presented as robust tools used to track global supply chains and also to compare production alternatives. However, although there are guide categories to date, the main indices and subindices of the indicators used have not been well established. In this way, it is assumed that the greatest problem encountered is actually the quantification of these social performances in the most diverse segments of the processes. Therefore, this is still a field to be explored in this approach. Thus, Table 5.2 shows some of the indices and indicators of social

TABLE 5.2 Indices, subindices, and indicators of social performance.

Index			
Subindex	**Indicator**	**Unit[a]**	**Equation/measure**
1. Health and safety			
Accidents Sick-leave	- Occupational accidents - Occupational fatal accidents - Sick-leave days - Preventive health measures	- Nr - Nr - Nr - Cat.	- Number of accidents per year per 1000 employees - Number of fatal accidents per year per 1000 employees - Sick-leave days per year per employee - Health measures (e.g., sick-leave analysis, health activities)
2. Adequate remuneration			
Payment Financial participation	- Payment according to basic wage - Average remuneration level - Capital participation - Profit-sharing and bonuses	- y/n - $ - y/n - y/n	- Payment of basic wage - Average payment per month per full-time employee - Existence of a capital participation model - Existence of a profit-sharing and bonuses model
3. Adequate working time			
Working time Work-life-balance	- Contractual working hours - Compensation for overtime - Access to flexible working time agreements - Rate of part-time employees	- h - Cat. - y/n - %	- Average contractual working hours per week per full-time employee - Compensation measures (e.g., exclusively payment, payment and free-time, exclusively free-time, any) - Access to flexible working time agreements (e.g., working time accounts, etc.) - Percentage of part-time employees per total employees
4. Employment			
Job conditions Duration of employment Job creation	- Rate of qualified employees	- % - % - % - %	- Percentage of employees with professional training per total employees

Continued

TABLE 5.2 Indices, subindices, and indicators of social performance—cont'd

Index			
Subindex	**Indicator**	**Unit**	**Equation/measure**
	- Rate of marginally employed - Rate of fixed-term employees - Rate of employees provided by temporary work agencies - Rate of recruitment	- %	- Percentage of employees earning maximum per total employees - Percentage of fixed-term employees per total employees - Percentage of employees provided by temporary work agencies per total employees - Percentage of new hired employees per year per total employees
5. Knowledge capital			
On-the-job training Vocational training Research and development	- Employees who participated in training - Support for professional qualification - Rate of vocational trainees - Rate of vocational trainees hired - Rate of employees in research and development	- % - y/n - % - % - %	- Percentage of employees who participated in training per total employees - Assumption of cost or exemption for training programs - Percentage of trainees per total employees - Percentage of trainees employed permanently per total trainees - Percentage of employees working permanently or temporarily in the research and development section per total employees
6. Equal opportunities			
Gender equality Older employees Minorities	- Rate of female employees in management positions - Measures to improve gender equality - Measures to support older employees	- % - Cat. - Cat. - % - %	- Percentage of female employees in management positions in relation to all employees in management positions - Measures for family support (e.g., support for child care, support for female employees) - Measures for older employees (e.g., offer of

TABLE 5.2 Indices, subindices, and indicators of social performance—cont'd

Index			
Subindex	**Indicator**	**Unit**	**Equation/measure**
	- Rate of disabled employees - Rate of foreign employees		part-time contracts, special equipment of the workplace) - Percentage of disabled employees per total employees - Percentage of foreign employees per total employees
7. Participation			
Workers' participation	- Works' council - Other measures for participation	- y/n - y/n	- Existence of works' councils in the organization - Measures to participate in the organization

[a] %, *percent;* Cat., *category;* h, *hours;* Nr, *number;* y/n, *yes and no.*
Based on Siebert, Bezama, O'Keeffe, and Thrän (2018).

performance, by way of exemplification, generally present in the social assessment of the life cycle.

Finally, as a factor of major awareness about the social life cycle assessment, it is the mode of obtaining the data, since this approach uses imprecise and site-specific data determined. In addition, since the measurement process can be quantitative, semiquantitative, or qualitative, this approach is complementary to the standard life cycle assessment and also supplementary to the life cycle cost assessment. Thus, we can apply it alone or in combination with other assessment techniques. Therefore, these results do not provide sufficient knowledge about a process or product against the decision to produce or validate the process. In this case, the acquired references of the social life cycle assessment are intended to provide a greater line of understanding for decision-making.

Life cycle costing

Traditionally, the life cycle cost (LCC) has been slightly defined as a methodology that enables cost comparisons over a specific period of time, taking into account relevant integral economic factors. That is, both initial costs and future

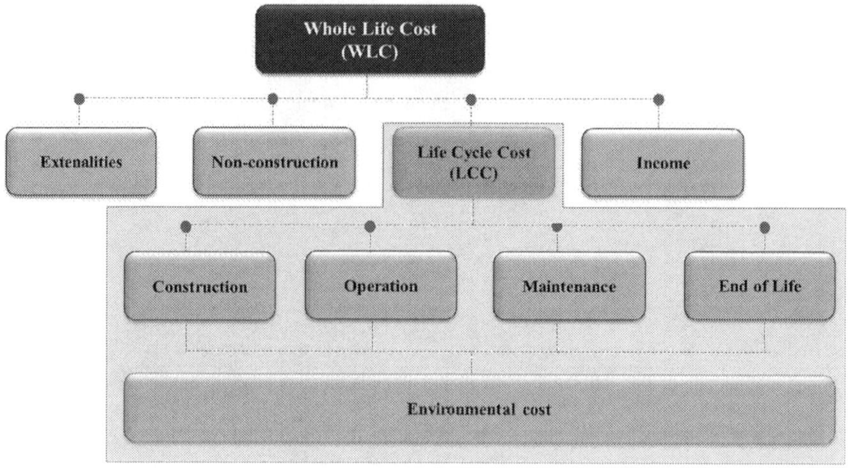

FIG. 5.3 Components of the whole life cost, the standard for life cycle cost environmental application.
This figure shows the different segments that are associated with the life cycle cost tool. In addition, it emphasizes that the whole cost of the process is also composed of other internal and external costs, not only the cost associated with environmental impacts.

operating costs were considered. In the past, therefore, the life cycle cost was applied to optimize product performance and lifetime cost of ownership.

More recently, different researches were developed to strengthen the LCC methodology for the industrial process and to use LCC in an environmental context. Thus, life cycle cost is now defined as the accounting for all internal and external costs associated with a process or product during its life cycle (Fig. 5.3). Thus, internal costs were established as costs incurred directly, i.e., capital, labor, energy costs, while external costs are those that are not incurred directly, i.e., resource depletion, water contamination, etc. While, external costs would have consequences for society in an environmental context beyond the limits of the process, but as a whole.

Similarly, the life cycle cost associates, through the specific categories of environmental impact defined in the standard life cycle analysis, as well as subcategories and indices of social performance structured in the social assessment life cycle, specific categories of accounting of the main costs involved in processes (Table 5.3).

Therefore, once the quantification categories have been defined, some basic life cycle costing equations have been established through the American Society of Testing and Materials (ASTM) publications (ASTM E917-17, 2017), being Eqs. (5.1)–(5.3):

$$PVLCC(t, \mathrm{i}, C) = \sum_{t=0}^{N} \frac{Ct}{(1+i)^t} \tag{5.1}$$

TABLE 5.3 Potential categories of simple life cycle cost assessment.

Conventional costs	Liability costs	Environmental costs
Capital equipment labor energy monitoring regulatory compliance documents maintenance insurance/special taxes Air emission control water effluent control radioactive/hazardous waste management raw materials/supplies waste treatment/disposal costs	Legal counsel penalties/ fines personal injury remediation activities economic loss property damage changes in future markets public image damage	Global warming Ozone depletion photochemical smog acid deposition Resource depletion water pollution chronic health effects Acute health effects habitat alteration Social welfare effect

$$PVCC = IC + PV(OM\&R + R + F - S) \tag{5.2}$$

$$i = (1 + r)(1 + I) - 1 \tag{5.3}$$

where *PVLCC* is the present value of the life cycle cost, *Ct* is the sum of all relevant costs per year *t*, *t* is the period of time, *N* is the duration of the study period, *i* is the nominal discount rate, *PVCC* is the main components of life cycle cost, *IC* is investment costs, *PV* is the present value (obtained through the nominal discount rate *i*), *OM&R* is operating, maintenance, and repair costs' nominal discount rate, *R* is replacement costs, *F* is depletion resource costs, *S* is the residual value, and *r* is the discount rate.

In this perspective, when using these mathematical tools and identifying the steps and analyzing the results generated, the role played by the life cycle costing can assume an economic-sustainable character when together with the life cycle assessment. Therefore, it results in a direct impact on the economic capital area of environmental protection and an indirect effect on economic prosperity.

Still, in the face of a relevant conception of life cycle costing assessments, this environmental cost analysis aims to help consumers make good decisions by encouraging producers to highlight a certain financial advantage of an environmentally sound product. This is because often environmentally sound products, identified by life cycle assessment, consistently have higher purchase costs but often become more accessible in the use or disposal phase.

Thus, assessment of environmental performance with the cost of life cycle becomes essential to assist the sustainability tripod. However, sound public policies against taxation become an emerging issue to globally improve future environmental performance and ultimately provide economically more viable and environmentally sustainable processes.

Life cycle sustainability assessment

In the current global context, paradigm shifts toward environmental protection have come to be seen as sustainable themes. In this way, sustainability is not only about environmental impacts, but also encompasses three dimensions such as the environment, social welfare, and economic parameters, aiming for balance as a whole.

Given this scenario, the United Nations Environment Program and the Environmental Chemistry and Toxicology Society have created a framework based on the development of a holistic life cycle sustainability assessment (LCSA). Thus, this approach to the sustainability tripod brings together tools such as life cycle assessment, life cycle cost, and social life cycle assessment, emerging as an alternative to process decision-making for more sustainable products throughout its life cycle (Fig. 5.4).

In addition, while sustainable development is feasible when the connection and interactions between the three dimensions are considered, some aspects of the consolidation of the life cycle sustainability assessment tool need further exploration. This is because, in the research status quo, the current applications of this tool are based on dependent analysis, not considering the interconnections of the assessments. Therefore, they use traditional individual methodologies, and only perform a sum of the results. However, this form of application results in negative and unintended consequences to the results, preventing the development of this tool.

As well as the numerous challenges already listed by standard life cycle analyses, social life cycle assessment, as well as life cycle cost, make life cycle sustainability assessment a tool at an immature stage of application. Due to this situation, some proposed studies insist that the union of the three pillars must be consistent and equivalent. In this way, the weighting becomes the highlight among the sequence steps of the life cycle sustainability assessment.

However, another bottleneck inherent in the weighting in this tool consists of at least two spheres, which are the weighting of individual indicators, i.e., the weighting applied to the individual impact categories and the weighting between the three dimensions of sustainability, e.g., the sum of the environmental results, and economic parameters in the same equivalence of punctuation. Therefore, the multicriteria evaluation scheme is strongly recommended because it presents appropriate results on aggregation resources.

Multicriteria analysis (MCA) has been applied very often in supportive circumstances in sustainable decision-making. In short, this analysis is applied in cases where there are coexistent evaluation criteria. In addition, this analysis

FIG. 5.4 Structure for life cycle sustainability assessment.
This figure represents the generic framework of life cycle sustainability assessment.

identifies goals and objectives and, subsequently, aims to identify the trade-offs. Therefore, its purpose is to incorporate qualitative and quantitative data into a common scope of the process.

Decision-making in a multicriteria analysis process usually includes steps such as problem definition, generation of alternatives and establishment of criteria, selection and weighting of criteria, evaluation matrix and sensitivity analysis. In this context, the main steps of multicriteria decision-making can be seen.

Defining the problem, generating alternatives, and establishing criteria

A decision-making problem begins by clearly determining the issue in question, and differing alternatives and scenarios. In addition, the actors, the objectives and any points in conflict, the degree of uncertainty, and the key issues should be indicated. Later, the problem can be framed indicating the evaluation criteria.

Assigning criteria weights

This step includes the assignment of criteria weights. These weights aim to present the relative importance of the criteria in the multicriteria problem. Frequently, these assignments are defined by techniques such as the analytical hierarchy process.

The analytical hierarchy process is a methodology that provides the user to determine punctual weights to select criteria, through a series of comparisons in pairs (Palma-Mendoza, 2014). This technique allows the aggregation of different impact categories into a single index, considering different amounts. It should be emphasized that usually when calculating weight scores at each hierarchical level, sets of weights are presented based on local, regional, or global perspectives. Namely, the environmental impacts are given certain amounts in relation to their geographical scale. In this sense, at the exemplification level, we can see in a global scope, more importance is attributed to global warming, while at regional scales the category of acidification and eutrophication receives greater amounts (Fig. 5.5).

Most Important	Criteria 1.1	Global warming
	Criteria 1.2	Acidification
	Criteria 1.3	Eutrophication
	Criteria 1.4	Ozone precursors
Least Important	Criteria 1.5	Human health

FIG. 5.5 Examples of the criteria for weight LCSA.
This figure illustrates some examples of weighting the impact categories, under hypothetical criteria of global, regional, or local scale.

In addition, the attribution of weights is performed by computational systems. However, because it is an affordable task, a calculator often becomes sufficient. Yet, the importance attributed by the criteria can be instructed by score rankings. In this way, simple techniques known as regular or ordinal rankings can be considered.

The regular ranking has as a characteristic to attribute, for each element relevant to the decision process, a certain score for a ranking associated with its importance. Structuring is indicated by 9-point scales (Fig. 5.6). Its main advantage is to specify the scores. However, as disadvantages, the scores are not satisfactorily discriminatory. In addition, the decision maker is able to choose the same assignments for different categories.

In contrast, ordinal ranking is defined as a technique capable of placing impact categories in order of importance (Fig. 5.7). In this ranking, each element is considered more or less important when compared to the other elements. In this way, ordinal rankings do not allow the same weights to be assigned to

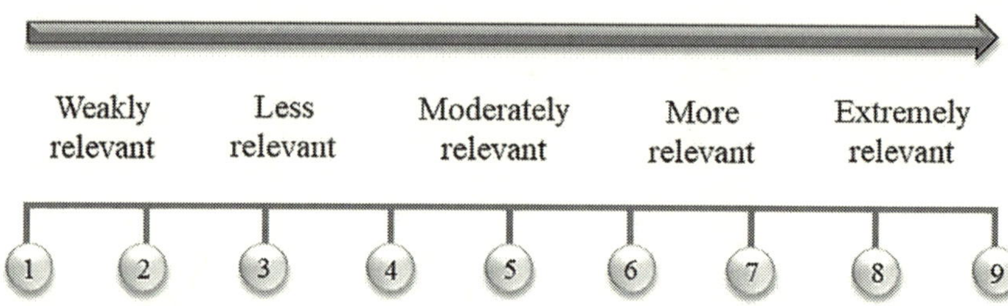

FIG. 5.6 Example for ranking regular scale.
This figure shows hypothetical weights attributed to the impact categories associated with a scale of greater or lesser importance in view of the environmental impact caused.

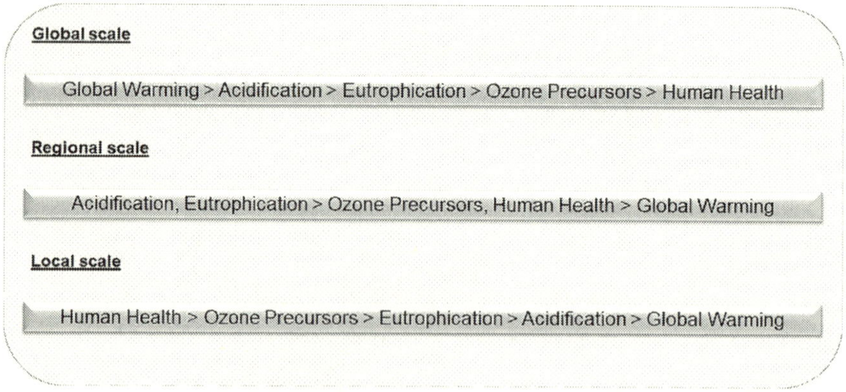

FIG. 5.7 Example of ranking ordinal for global scale.
This figure shows a practical example applied to weighting on a global scale, which considers impacts from minor to major importance.

different categories. Thus, among the advantages, it should be emphasized that this ranking has no ambiguities and is capable of substantial discrimination of the degree of importance between the categories.

Subsequently, after the attributions of weights to the criteria, it is necessary to apply the general score of the values. This occurs when the scores are weighted. In summary, the general preference score is given by Eq. (5.4):

$$S_i = \sum_{j=1}^{n} w_j S_{ij} \tag{5.4}$$

where S_i is the general preference score, w_j is the criterion weight, and S_{ij} is the criteria of the overall score for option j.

In other words, it is essential to multiply the score of an option by a criterion by the weight of importance of the criterion. It is important that you do this for all criteria, then subsequently add the products to give the overall preference score for that option. Then repeat the process for the remaining options.

Construction of the evaluation matrix

This phase is characterized by elaborating the model that constitutes the evaluation process, based on the problem. Thus, the elaborate complex image is extracted, so that the problem can be properly evaluated. At the end of this step, the problem of multicriteria analysis can be expressed in matrix form as follows:

$$\begin{matrix} A_1 \\ A_2 \\ \cdot \\ \cdot \\ \cdot \\ A_M \end{matrix} \begin{bmatrix} X_{11}\ldots & X_{1n} \\ X_{21}\ldots & X_{2n} \\ \cdot & \cdot \\ \cdot & \cdot \\ \cdot & \cdot \\ X_{m1}\ldots & X_{mn} \end{bmatrix}$$

where A is the evaluation given to alternative with respect to the criterion, n is the number of criteria, and m is the number of alternatives.

After, the matrix is structured, the established values of the cells are divided by the total sum of the column. Thus, the values resulting from the summation per line are divided by the total number of impact categories, resulting in the final weights (Eq. 5.5):

$$V_{ij} = \sum_{i=1}^{i} \left(\frac{CV}{CT \times i} \right) \tag{5.5}$$

where V_{ij} is resulting in the final weights, CV is the cell value of matrix, CT is the total sum of the scores assigned to the impact categories, and i is the category of impact.

Subsequently, potential impacts (C_i) are normalized (N_i), through specific factors as seen in Chapter 4, and multiplied by an evaluation factor (V_{ij}). They are still added to result in the environmental index (I_j) (Eq. 5.6):

$$I_j = \sum_{i=1}^{Z} \left(\frac{C_i}{N_k} \times V_{ij} \right) \tag{5.6}$$

where Ij is the final environmental index, Z is the sum of the impact categories i, C_i is the potential impacts, N_k is the normalization, and V_{ij} is resulting in the final weights.

Sensitivity analysis

Indices obtained are strongly encouraged to undergo a sensitivity analysis. This step aims to assess whether the preferences or weights assigned to the impact categories affect the overall order of the results. Therefore, it is suggested that other scenarios be elaborated, so that the results can be compared. In addition, this sensitive strategy may help to improve the scenarios originally considered. Thus, performing the above steps applied to industrial production processes becomes a potential tool to establish the ideal models of sustainable processes.

References and recommended reading

Papers of particular interest have been highlighted as:
• of special interest
•• of outstanding interest

Andrews, F. M., & Withey, S. B. (2012). *Social indicators of well-being: Americans' perceptions of life quality*. Springer Science & Business Media.

• ASTM E917-17 (2017). *Standard practice for measuring life-cycle costs of buildings and building systems*. West Conshohocken, PA: ASTM International This international standard has been developed in accordance with the internationally recognized principles of standardization set out in the World Trade Organization's Committee on Technical Barriers to Trade (TBT) Decision on the Principles for the Development of International Standards, Guidelines and Recommendations, to define life cycle cost methods suitable for determining the costs of systems and products.

Palma-Mendoza, J. A. (2014). Analytical hierarchy process and SCOR model to support supply chain re-design. *International Journal of Information Management*, *34*(5), 634–638.

Siebert, A., Bezama, A., O'Keeffe, S., & Thrän, D. (2018). Social life cycle assessment indices and indicators to monitor the social implications of wood-based products. *Journal of Cleaner Production*, *172*, 4074–4084.

Chapter 6

Carbon footprint and carbon market

Originally, the exponential increase of atmospheric emissions coupled with concerns about the global population has spurred a need to quantify and express total greenhouse gas emissions. In a brief historical context, the emergence of the carbon footprint terminology was based on the methodology known as the ecological footprint and emerged as an alternative to characterize carbon quantifications in a peculiar way (Wackernagel & Rees, 1996).

More specifically, the carbon footprint was defined as the biologically productive area required to sequester sufficient carbon molecules to prevent the increase of atmospheric carbon dioxide. Today, the most practical definition of carbon footprint is determined by the net exchange of greenhouse gases per unit of product or process. Therefore, many researchers believe that the best way to quantify these emissions is through the broad scope that the life cycle assessment tool provides. However, the independence gained by this new methodology has become widely used and is known as an indicator of global warming potential (Finkbeiner, 2009). Yet, on a more contemporary look, the carbon footprint has been defined on different system perspectives and limitations, including terminologies such as carbon content, embedded carbon, carbon flows, virtual carbon, greenhouse gas footprint, and climate footprint (Table 6.1) (Wiedmann & Minx, 2008).

However, there are different concepts on carbon footprint, the absence of robust methods for quantifying direct and indirect emissions from the embodied carbon flows has resulted in persistent misconceptions throughout the quantitative design of carbon determination processes.

This way, how to measure, include, and present the carbon footprint?

Carbon footprint assessment standards

The carbon footprint refers to the direct and indirect impacts of carbon dioxide emissions produced within the boundaries defined by industrial processes, companies, or projects. The products resulting from this assessment may be exclusively concentrated in carbon emission inventories. In addition, information on

Sustainability Metrics and Indicators of Environmental Impact. https://doi.org/10.1016/B978-0-12-823411-2.00009-8

TABLE 6.1 Concepts on carbon footprint found under different perspectives in the literature.

Concept	Indicator	Reference
Carbon footprint is defined as the amount of carbon emitted anthropogenically	Social life cycle assessment	BP (2007)
The carbon footprint is defined by measuring carbon dioxide equivalent emissions from commercial facilities	Social life cycle assessment	British Sky Broadcasting (Sky) (Patel, 2006)
The carbon footprint is estimated by quantifying greenhouse gases equivalent to the production of raw material used in its manufacture, to disposal of the finished product	Product life cycle assessment	Carbon Trust (2007)
The carbon footprint encompasses the quantification of direct and indirect carbon dioxide equivalent emissions from process and use activities	Social and Product life cycle assessment	Energetics (2007)
Carbon footprint is defined as monitoring the impact of human activities on the environment in terms of carbon dioxide equivalents	Environmental life cycle assessment	ETAP (2007)
The carbon footprint is defined by the ability to biosequester carbon dioxide emitted by anthropogenic activities	Environmental life cycle assessment	Global Footprint Network (GFN, 2007)
The carbon footprint is defined as a way to quantify carbon dioxide emissions emitted through the combustion of fossil resources. In a more specific context, within an organizational structure, it is the amount of carbon dioxide emitted as a result of operating activities generally related to energy represented by the product or process in question	Life cycle energy assessment	Grubb and Ellis (2007)
Carbon footprint is defined as the total portion of greenhouse gas emissions emitted over the entire life cycle of a process or product	Life cycle assessment	Parliamentary Office of Science and Technology (POST, 2006)

Based on Wiedmann and Minx (2008).

greenhouse gas emissions can be described in full carbon footprint reports designed for public organizational institutions. However, there is now a consensus that the standardization of carbon footprint assessment methods is based on environmental burdens related to process input and output flows. The main step standardizations to consider when calculating a carbon footprint are shown in Fig. 6.1.

Regardless of industrial tracking, process engineering is based on inbound and outbound flows, and therefore metaphorically can be approached as a river. For a better understanding, the terminologies used to describe upstream and downstream processes in the carbon footprint diverge from the concepts already established in engineering regimes. As a well-known, in its usual short definition, upstream refers to all unit operations required to group together the materials needed to create a product. In this case, the upstream stage is characterized as the search and extraction of raw materials without its processing. In contrast, downstream processing encompasses the processing of raw materials collected during the upstream stage, resulting in the finished product.

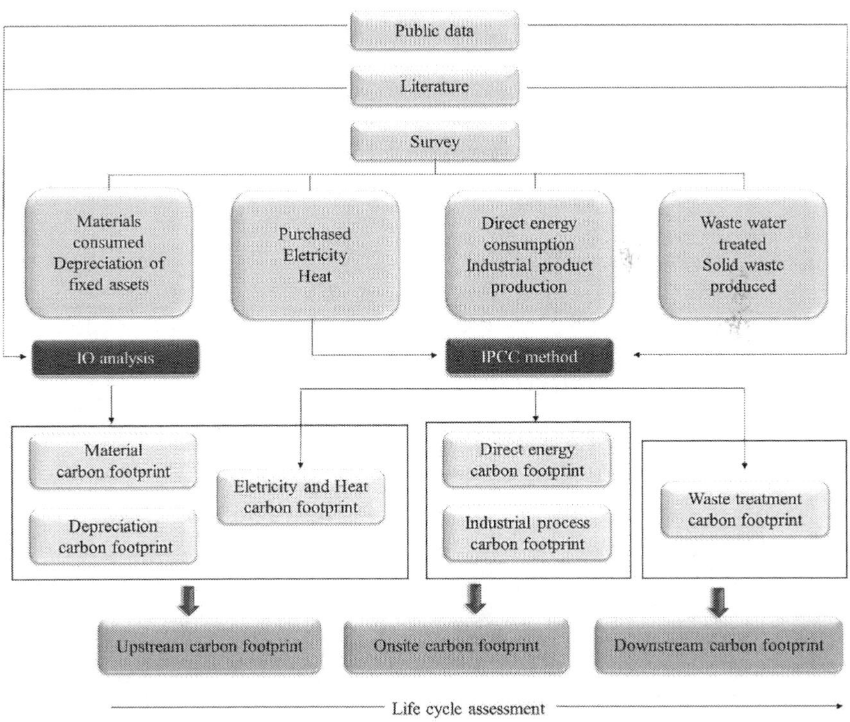

FIG. 6.1 Framework approach to carbon footprint determination.
This figure aims to elucidate the two main methodologies for quantifying the carbon footprint, considering the system's input and output analysis and the IPCC method. *(Based on Dong, Geng, Xi, & Fujita, 2013.)*

On the other hand, when we refer to downstream and upstream processes in the carbon footprint, we have to take into consideration that upstream processes of the carbon footprint refer to the emissions caused during the production or process of a product. While downstream carbon footprint processes have been substantiated as quantifications of greenhouse gas emissions generated by a product or service when they are used and disposed of by consumers. At the level of exemplification, the National Environmental Policy Act (EUA) understands as downstream processes of the proposed action, such as the burning of fossil fuels, while emissions from activities that occur "upstream" of the proposed action would be as the extraction of fuel fossils. Thus, from a carbon footprint point of view, "upstream" can mean that polluters and downstream can mean consumers.

However, it is important to note that upstream and downstream processes depend heavily on the processing plant as well as on the limits of the systems defined in the early life cycle assessment. Therefore, deciding which part of the value chain to focus on is a challenging task, as understanding the indirect emissions in your value chain often results in unpredictable exposures. In addition, the management of indirect emissions associated with the use phase needs to be evaluated accurately and rigorously, and partnering with a specialist with proven carbon management expertise is a logical first step in reducing value chain carbon emissions.

In view of this, the management of upstream carbon regulations would have two main advantages over a downstream approach. At first, a higher percentage would be fully covered against total emissions. And in a second moment, the ease of conducting accounting would increase as fewer regulatory sources would manage it.

Thus, some specific criteria must be followed and protocoled in order to ensure the reliability of quantified results through carbon footprint, being them the greenhouse gas selection, the methodology procedure of evaluation of carbon footprint, methodological applications, the definition of system delimitations, and use of mathematical model appropriate.

Selection of greenhouse gases

Numerous methodologies present opinions on the gases that should be accounted for in the carbon footprint. Among the gas reduction efforts, the most rigid approach is from the point of view of the Intergovernmental Panel on Climate Change, as it requires the quantification of the main gases, i.e., carbon dioxide, methane, nitrous oxide, hydrofluorocarbons (HFCs), perfluorocarbons (PFCs), and SF 6. By contrast, the European Emissions Trading Scheme only asks for a description of CO_2 emissions (Ellerman & Buchner, 2007). In addition, some authors suggest using the restricted carbon footprint term as the exclusive carbon dioxide-based assessment. They claim that there are promising advantages, as carbon dioxide measurement is relatively affordable compared to other gases. However, it is necessary to be aware of the risks of oversimplifying quantifications as the end result may lead to inaccurate responses. Thus, these different accounting conceptions establish a significant disorder regarding the inclusion or exclusion of greenhouse gases and the carbon footprint.

Conversely, accounting for CO_2 emissions alone would estimate about two-thirds of greenhouse gases. In addition, the omission of CH_4 would result in the exclusion of almost one-fifth of greenhouse gases and a significant gap in their overall management. Thus, an alternative to these shortcomings in carbon footprint orientation and practice can be remedied through the global warming potentials of greenhouse gases. As seen in Chapter 3, global warming potential is the mensuration of greenhouse gases that contribute to global warming over a time horizon and compares each gas with the same amount of carbon dioxide and its reference potential (Annex 1). Thus, the way carbon protocols are presented to decision-makers is a crucial factor in establishing a beneficial and consistent carbon footprint. Therefore, the most common carbon measurement metrics are by the equivalence of other greenhouse gases to CO_2 equivalents.

The methodology procedure of evaluation of carbon footprint

From a methodological perspective, the carbon footprint has been presenting several environmental calculators as well as international criteria. However, to date, the methods used to establish the carbon footprint are not yet specified. Given this aspect, quantification methodologies require basic requirements that satisfactorily meet the assumptions of the definition. Thus, a carbon footprint can be analyzed for different units at different scales and using different methods.

Currently, the three main methods of quantifying carbon emissions consistent are input-output analysis (IO), life cycle assessment (LCA), and hybrid analysis (IO-LCA). In short, as the basis of the life cycle assessment methodology, all carbon accounting methods require a functional unit via scale (Fig. 6.2). From a more practical point of view, researchers and consultants

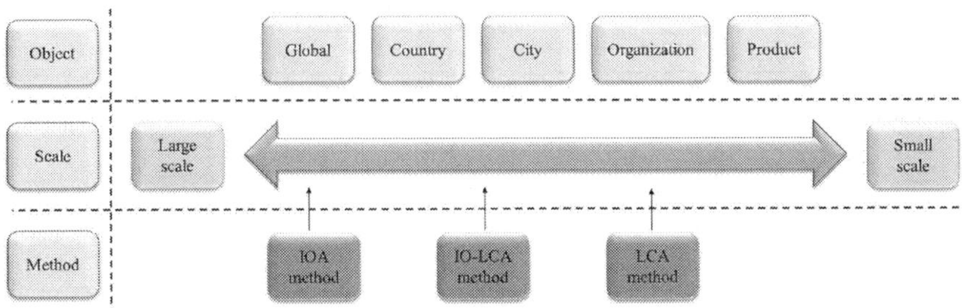

FIG. 6.2 Outworking and methodologies for the carbon footprint.
This figure shows that regardless of the methodology used to quantify the carbon footprint, it is necessary that the analyst consider the functional unit as the main parameter of the assessment.

suggest that the application of the life-cycle assessment methodology is better suited to consumer products, while nationwide applications would opt for input and output analysis (Gao, Liu, & Wang, 2013).

In addition, the bottom-up method involves a process analysis considering the environmental impacts of individual products from the cradle to the grave. However, this will have a system boundary problem only onsite, emphasizing the need to identify appropriate system boundaries. Life cycle assessment based on process analysis encounters difficulties for larger entities, such as government, social institutions, or specific industrial sectors. Conversely, the hybrid methodology, i.e., combining the strength of both life cycle assessment and input/output assessment or processes analyzes, emerges as an active research alternative and is increasingly being used in practice. Moreover, this methodology ensures the robustness and accuracy of each method, providing an increase in the overall reliability of the estimates.

Therefore, inherent in life cycle assessment, when applying these carbon calculation methodologies, specific procedures regarding the definition of methodological applications and system boundaries need to be established in order to identify in detail any flows included in processes or facilities.

Methodological applications

The correct application of carbon footprint accounting methodologies is proposed in order to increase consistency and ensure comparability of results between different application levels. This becomes even more important once a common standard for footprint accounting is reached.

Traditionally, the primary interest in carbon footprint has emerged with the need to quantify product carbon footprints (PCFs). Understanding the results achieved aspired to improve the environmental performance of the product or service, as well as encouraging consumers to buy more sustainable products. However, bottlenecks in the face of difficulties in collecting data, high time demands and lack of methodological consensus, drastically reduced the initial euphoria.

Over the years, methodological advances gained have spurred the development of different structures and inventories to make carbon footprint procedures less confusing. Thus, the carbon footprint has expanded to nations, regions, and cities. However, although the interest granted, fewer studies have been reported in view of these system boundaries. Governed by the Intergovernmental Panel on Climate Change, participating countries are required to submit an annual statement of all greenhouse gas emissions associated with all activities within their boundaries. Thus, as nations' carbon footprints are closely linked to debates about equity in allocating emissions, production accounting can lead to major underestimations of carbon emissions from products and services. As such, a large proportion of carbon emissions come from supply chain processes. Therefore, local and regional governments are responsible for many decisions that affect greenhouse gas emissions, such as transportation and land use planning, zoning and building standards, and the management of their own activities.

Alternatively, the carbon footprint in the organizational context was created with the objective of sectioning processes to facilitate the identification and quantification of carbon emissions for business units or municipal organizations. Nowadays, the most accepted methodology is the greenhouse gas protocol developed by the World Resource Institute and the World Business Council for Sustainable Development. However, the demand for an internationally recognized standard and consistency in the calculation of organizational greenhouse gas inventory reports has led to the creation of an international organization for standardization 14064 (ISO, 2006).

In recent years, efforts toward an environmentally sound lifestyle have spurred the development of personal carbon footprints. This tool raises awareness among the population by providing calculators that account for the greenhouse gas emissions associated with their routines (Wintergreen & Delaney, 2007). In some cases, they occasionally suggest requirements and alternatives include normal activities to mitigate or effect carbon offsets. In contrast, inconsistencies attributed to the different available virtual calculators may come in results with carbon tonnage deviations.

Therefore, without a fair, equitable, and accurate emission allocation method, carbon trading or taxation will be impossible.

In short, Table 6.2 presents the different classifications that can be applied for the carbon footprint.

TABLE 6.2 Methodologies and their respective applications of carbon footprint.

Methodology	Enforcement
Determination of a carbon footprint of a territory or country	Applied for the purpose of drawing up local emission inventories
Quantification of organizational carbon footprints or application of corporate analytics	Detect emissions from all activities throughout the organization, including the energy use of facilities, industrial processes, and vehicles
Assessment of the carbon footprint of a product or service	Identify emissions over the entire life of a product or service, from raw material extraction and manufacturing to its use and reuse, recycling, or final disposal
Calculation of the carbon footprint for personal use	It consists of the quantification of social emissions associated with individuals and their respective activities

System delimitations

As with life cycle assessment, system boundaries must be specified in detail. Regardless of the methodological application, the carbon footprint has two extra concepts of system boundaries, namely organizational boundaries and operational boundaries. Therefore, the principles for choosing methods need to be grounded in this understanding.

The organizational limit refers to the established boundary that is directly related to the particularities of the process. This, in turn, should be selected to portray the industrial system based on financial or corporate management. In some cases, management may assume the responsibilities of issues, which are called equity interests. In view of this, the operational limits for the selection of direct and indirect emissions, which will be accounted for, should be established in order to facilitate quantification as well as to restrict the levels or scopes that will be evaluated (Carbon Trust, 2007).

Besides, as a second definition, operational limits can be described as direct emissions emitted on-site and elements associated with indirect emissions, i.e., emissions which are included during energy consumption. However, indirect emissions as a whole are characterized by being complex to quantify as they constitute one of the largest segments within industrial processes.

In general, the system boundaries are divided into process steps (Fig. 6.3). In short, these steps are determined in order to separate direct emissions, known as local emissions, from emissions incorporated into energy and also from all indirect emissions, which were not included in the previous steps.

At the level of exemplification, Fig. 6.3 addresses the estimated carbon footprint steps. As previously seen, steps II and III include direct emissions, however, step II portrays emissions from power generation or purchase, transmission, and distribution originated by the process under assessment. Also, note that end-use emissions are not considered in step II. In addition, step III encompasses all other emissions entered within the specified limit. In contrast, it is noteworthy that stage III has been slightly determined, as most studies are limited to stage II, as a result of the complexity of accurately quantifying beyond stage II. Because of these quantification bottlenecks, much of the standardization of greenhouse gas counting enables step III to become optional.

Additionally, while care is taken to delimit systems and to choose the best method, the accuracy of carbon footprint calculations is directly related to the collection of consumption data required for all emission sources within the given boundary. Besides, it is relevant to clarify any gaps in the data and to list any assumptions that were made in the footprint calculation.

In this context, current trends in terms of several publications, targeting of research, government demand, and the generation of increasing organized data on effect gas flows have spurred the development of mathematical models to guide research carbon footprint quantifications. In addition, mathematical

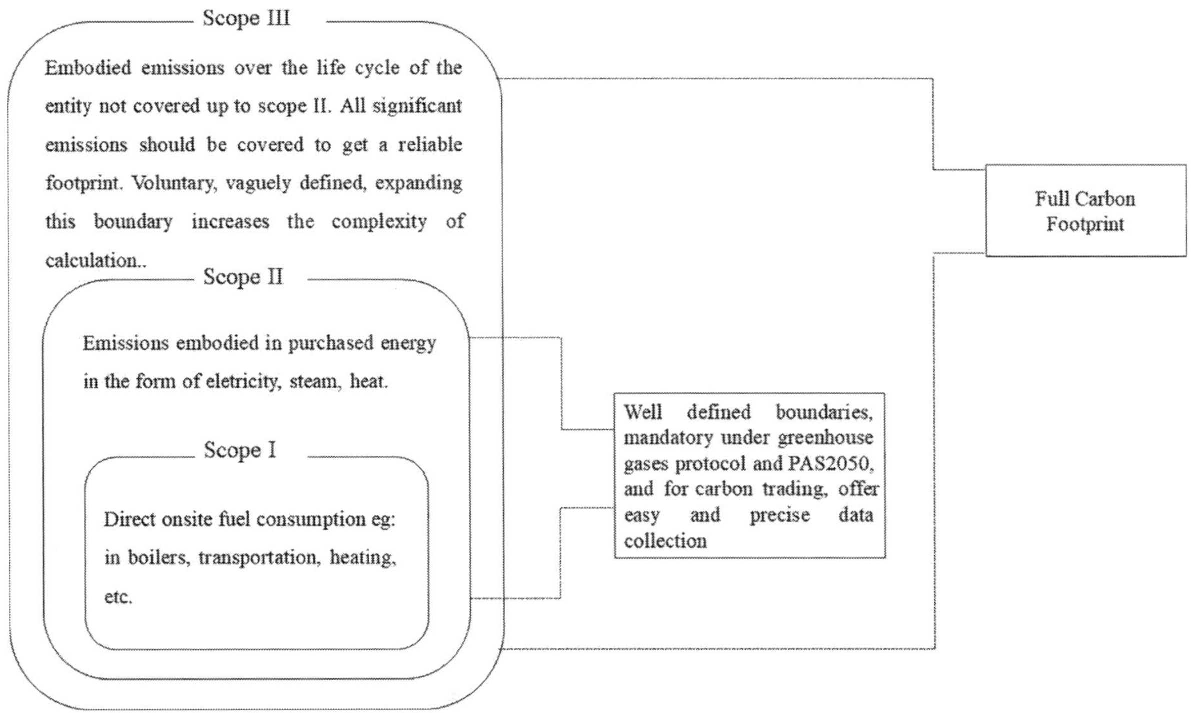

FIG. 6.3 Boundaries for carbon footprint quantification.
This figure presents as a hypothetical model of how the limits of the system applied to the quantification of the carbon footprint can be divided. It is important to highlight the division of direct and indirect emissions associated with the system under assessment.

modeling is an indispensable tool for the generation of greenhouse gas balance estimates and the assessment of mitigation strategies.

Mathematical modeling

Onsite carbon footprint

The internal carbon footprint is quantified by the sum of the carbon footprint of direct energy consumption (greenhouse gas emissions resulting from the combustion of fossil resources within a given industrial establishment) and the carbon footprint required for processing (greenhouse gas emissions generated through direct chemical or physical conversion of materials within plant processes). The process of calculating the carbon footprint of direct energy consumption is shown in the following equations:

$$CF_{energy}^{GHG_k} = \sum_{i=1}^{n} M_i \times EF_{i,GHG_k} \tag{6.1}$$

$$CF_{energy} = \sum_{k} CF_{energy}^{GHG_k} \times GWP_k \tag{6.2}$$

where represents emissions of GHG_k caused by direct energy combustion (tons); k represents the type of different GHG (kg eq of CO_2, CH_4, N_2O, SF_6); CF_{energy} represents carbon footprint of direct energy consumption (ton CO_2eq); M_i represents total consumption amount of fuel i (ton); i represents the type of energy fuel; $EF_{i,GHG}$ represents emission factors of GHG_k of fuel i (ton GHG_i/ton fuel); and GWP_k is the global warming potential of GHG_k.

Likewise, the process of quantifying the carbon footprint of industrial processes is supported and declared by the intergovernmental panel on climate change, known as national greenhouse gas inventories, which gave rise to the following fundamental equation:

$$CF_{process}^{j} = P_j \times EF_j \times GWP_k \tag{6.3}$$

where, is the direct carbon footprint of industrial process j (ton CO_2eq); P_j is the weight of product j (ton); and EF_j is the CO_2 emission factor of industrial product j (ton CO_2/ton product).

Upstream carbon footprints

Upstream carbon footprints include electricity purchased and thermal carbon footprint, material carbon footprint, and depreciation carbon footprint. Most industrial installations would originate from external electricity demand and heat supply, except where there are own heating projects such as the use of power plants and boilers. Given this aspect, to prevent double quantification, only acquired electricity and heat from outside are considered. The quantification of this portion of the carbon footprint is similar to the carbon footprint of direct energy, i.e., equal to the value of electricity demand and heat multiplied by their respective emission factors. However, it is important to note that the scope of manufacture and supply of electricity and heat is one of a particular degree because it not only consumes electricity and heat, but also supplies mainly electricity and heat to the site. Thus, the carbon footprint of electricity and heat purchased is equal to consuming the carbon footprint of electricity and heat less the carbon footprint of electricity and heat supplied.

The carbon footprint of upstream materials is novel and should be calculated based upon input/output assessment. For each industrial follow-up, the carbon footprint of upstream materials should be calculated according to

$$\varepsilon_{mat} = \sum_{k} E_k \times V_{mat,k} \tag{6.4}$$

where ε_{mat} is the carbon footprint of upstream materials consumption of the studied company (ton CO_2eq); $V_{mat,k}$ is the total consumption of material k by the company; and E_k is the embodied emission intensity of the industrial sector to which this material belongs to (ton CO_2eq/company).

Sector classification is based upon the Classification of National Economy Industry. The embodied emission intensities of different economy sectors (E) can be calculated by

$$E = D(I - A)^{-1} \tag{6.5}$$

where E represents the vector of embodied emission intensity; D represents the direct emission intensity of the industrial sector; I represents the identity matrix; and A represents the intermediate demand matrix and describes the relationship between all sectors of the economy.

Additionally, the depreciation carbon footprint is similar to that of the materials. The difference is that the value is the total depreciation value of each enterprise rather than material consumption value. It can be calculated through Eq. (6.6):

$$\varepsilon_{dep} = E \times V_{dep} \tag{6.6}$$

where ε_{dep} represents the depreciation carbon footprint of the studied company (ton CO_2eq); V_{dep} represents the value of current year depreciation of the studied company; and E represents the embodied emission intensity of the industrial sector to which this company belongs to (ton CO_2eq/company).

Similar to carbon footprint of materials, depreciation carbon footprint of industrial sectors and the whole industrial follow-up can be obtained based on the sums of carbon footprints of all the studied companies.

Downstream carbon footprint

The downstream carbon footprint refers to the greenhouse gas emissions caused by the waste treatment step, comprising the treatment of air pollutants, wastewater treatment, and solid waste treatment. In addition, in a more careful analysis, solid waste generated by human activities within the industrial facility is considered. However, it is known that the total amount of this industrial waste is very small and can be neglected if the data are not available or is very marginal. For wastewater treatment, emissions of CH_4 and N_2O should be considered. Thus, Eqs. (6.7) and (6.8) show how to calculate such values, respectively.

$$E_{CH_4} = (TOW - S)EF - R \tag{6.7}$$

where E_{CH_4} represents CH_4 emissions generated during industrial wastewater treatment process (ton CH_4/yr); *TOW* represents the total degradable material in the wastewater (ton COD/yr); *S* represents the organic component removed as sludge (ton COD/yr); *EF* represents the emission factor of CH_4, with a default value of 0.025 t CH_4/t COD; and *R* is the amount of CH_4 recovered in the inventory year (ton CH_4/yr).

$$E_{N_2O} = N_{IW} \times EF_{IW} \times \frac{44}{28} \tag{6.8}$$

where E_{N_2O} represents N_2O emissions generated during industrial wastewater treatment process (ton N_2O/yr); N_{IW} represents the total weight of nitrogen in the effluent (ton N/yr); EF_{IW} is the emission factor for N_2O emissions of wastewater, with a default value of 0.005 ton N_2O-N/ton N.

Finally, when reporting and verifying the data collected on carbon footprints inside facilities, processes and products, organizations should prepare a report to facilitate inventory verification to inform external or internal users and also enhance the visualization of these results by directing them to be included in national or international greenhouse gas management programs. However, as with the standard life cycle assessment, a third-party verification of the carbon footprint is suggested to be undertaken to add credibility and confidence to the carbon report for public disclosure.

Carbon certifications

Strategies to reduce greenhouse gas emissions, particularly the carbon footprint, have become a global consensus to respond and mitigate global climate change. Ideally, as initiatives set out to reduce their energy demands as well as carbon footprints, efforts were made to return these measures. As a result, international certification seals are designed to recognize the competence of industries, services, and individuals to quantify, evaluate and report their emissions.

Several certification companies are scattered around the world, with the primary objective of ensuring that projects reduce emissions and promote sustainable development. However, some concepts about how to name certain carbon reduction actions have resulted in the conception of numerous terms associated with carbon footprint neutrality and climate positivity/carbon negativity (Table 6.3).

TABLE 6.3 Definitions and application concepts about carbon footprint terminology.

Terminology	Concept
Carbon neutral	It means that an activity releases zero carbon emissions into the atmosphere
Negative carbon or positive climate	It expresses that an activity is going to attend zero liquid carbon emissions to really raise an environmental benefit by removing additional carbon dioxide from the atmosphere
Positive carbon	It is primarily a marketing term and is defined as process surplus renewable energy

In a brief explanation, these definitions fully share the same latent purpose, that is, the substantial neutralization of the carbon footprint of a product or process. In most cases, the terms zero carbon or carbon neutral are applied to certify installations or industrial products that use renewable energy sources in the establishment. In addition, for a predetermined period of time, usually 1 year, the amount of net energy generated for the system is equivalent to the required net energy (Fig. 6.4). On the other hand, positive carbon stands out by the step forward in relation to neutral carbon and zero carbon, because it consists of making additional contributions, thus the installations produce local renewable energy and these feed the required demand of the installation and at the same time can be exported to other locations, thus compensating the energy demand of other installations. Therefore, while zero carbon is reputed as the benchmark for current processes and products, positive carbon certifications will play a

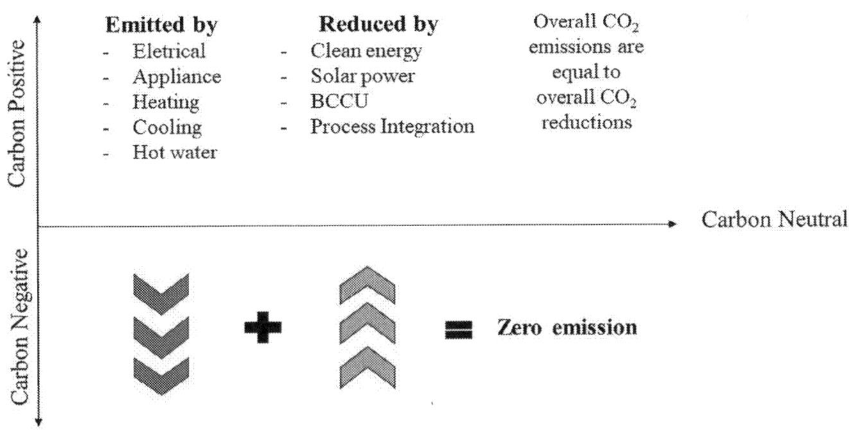

FIG. 6.4 The global approach to carbon neutrality.
This figure illustrates the differences among a system considered positive, neutral, or negative carbon.

promising role in the future of industrial facilities to limit climate change. To this end, the importance of applying the carbon footprint methodology framework is critical to any management initiative. Thus, members responsible for industry administration generally seek the assistance of certification companies which are qualified to implement the carbon footprint methodology.

In the current state of certification, groups of independent professionals unite with a commitment to accelerate the transition to a sustainable low-carbon economy. Recognized globally, initially established by the British government in 2001, Carbon Trust was developed in this format to provide advice to a wide range of audiences, be they governments, businesses, and public sectors expanding smart policies through strategies for utilizing resources seeking greater efficiency, sustainability, and credibility in their processes. In addition, innovative solutions using versatile clean technology investment alternatives aimed at consolidating cutting-edge solutions in the market are proposed.

Similarly, founded in 2001, the Department for Environment Food and Rural Affairs (DEFRA) presents a broader mission as a proposal for sustainable development. In addition, DEFRA has strategic priorities, which include climate change and energy, sustainable consumption and production, protection of natural resources, and sustainable rural communities.

Most recently, produced and published in April 2010 by the British Standards Institution (BSI), along with organizations such as the Carbon Trust and DEFRA, the PAS 2060 carbon neutrality standard came into effect. In short, this certification seal aims to specify the requirements to be met by an organization seeking to demonstrate carbon neutrality by quantifying, reducing, and offsetting greenhouse gas emissions.

Likewise, governed by ISO 14043, SCS Global Services is also another certification company that voluntarily regulatory guides carbon footprint verification for industrial, nonprofit, corporate, and municipal customers, including aviation, waste management, manufacturing, transportation, high technology, and professional services.

Finally, while certificates are an attractive alternative to publicizing greenhouse gas mitigation, there is a need for policies that specifically encourage homeowners such as tax exemption. In this context, the actual impact assessments of the processes will, therefore, be critical to the cost-benefit assessment of policies to encourage green certification.

Market carbon

Although much has been reported and commented on the carbon footprint in industrial processes and systems in internal audits and international conferences often related to the environmental footprint, only recently has increased attention been paid to this area in the face of economic measures. Thus, in recent years, the carbon market has emerged as an important indicator related to carbon management and its respective economic potentials in order to promote the reduction of carbon emissions through low cost.

Before we can access the carbon market structures and how they operate, we need to be aware of their definition. Therefore, a carbon market is widely known as carbon emissions trading. These negotiations were created through a common agreement between countries participating in the Paris agreement, in order to reach the pacts dealt with in the face of greenhouse gas emission reductions.

The carbon market has a structure that is quite easy to understand and is followed by two main categories, this is the quota market and the voluntary trading market. In short, the quota trading market is willing to focus on the reduction targets of released gases (Hawkins, 2016). In contrast, the voluntary trading market, as the name implies, is presented through voluntary transactions, being a social commitment. In addition, there is a second subdivision of the quota market, which is based on two further categories, namely the quota-based market and the project-based market (Fig. 6.5) (Rodriguez-Lopez, Engels, & Knoll, 2017).

Carbon markets, in turn, have been presenting potential outcomes against climate change policies. Key roles in sustaining markets are drawn from key statements issued by industry sectors and are hierarchized across national and/or global policies, regulatory and institutional frameworks.

Notably, a number of policymakers and entrepreneurs emphasize the continuing struggle to mitigate greenhouse gases through attempts that correlate with economic issues. In addition, information processors underscore the interest in pricing carbon as part of their plan to drive changes in process, production, and consumption patterns, as well as directing investments toward low-carbon options.

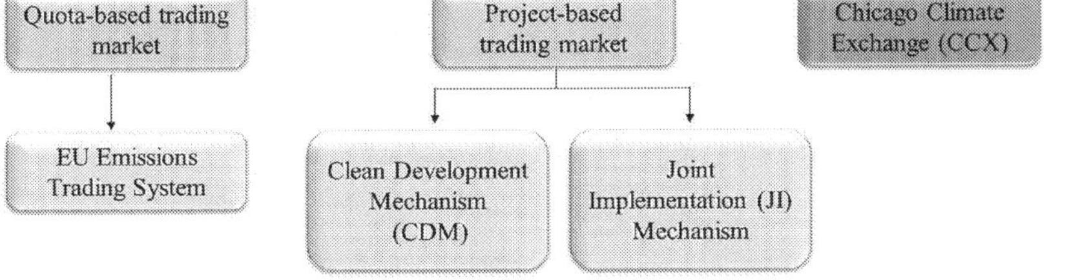

FIG. 6.5 Global carbon market structure.
This figure illustrates the main divisions of the global carbon market.

Given these aspects, carbon markets play indispensable tools in the progress of managing the future of climate change. Thus, carbon pricing becomes a valuable resource for manipulation against policy strategies. In addition, some government strategies exploit priority mechanisms by designing policies and structures to raise investor interests for low-carbon processes. Therefore, the carbon market represents and consolidates a probable and practical regulatory and institutional organization.

Among the main benefits of carbon markets, we can highlight the series of efforts to achieve the reduction of greenhouse gases. However, other more cost-effective privileges for mitigations can also be seen. In addition, carbon markets substantially encourage modern engineering applications, which require less demand for energy resources. In common sense, their existence helps in the gradual subtraction of current carbon costs. Therefore, carbon pricing mechanisms act as market drivers that aim to achieve reduction targets at a minimum carbon cost.

In terms of competitiveness, carbon markets develop resistance to negative external situations caused by greenhouse gas emissions by assessing and weighing the real cost of externalities related to different technologies. Thus, these managements should play attractive roles vis-à-vis managers and provide adequate comparisons. Thus, carbon markets must offer equal conditions between institutions and corporations through mechanisms that promote private sector solutions.

In contrast, cap-and-trade (CAT) schemes emerge as a potential alternative to standardize carbon pricing through emissions. CAT schemes are currently the most affordable method of regulating carbon dioxide and other emissions. Recurring scheme managers assume that they set a certain ceiling for agreed emissions. Subsequently, distribution or auctioning occurs in order to release emissions allowances that total the predefined limits. What may happen, however, is that institutions that do not have sufficient licenses to fund their emissions are subject to reductions or boosted by extra credits from another company (see "Carbon credits" section).

However, carbon markets and carbon taxes share many of the same theoretical properties and may be identical in many respects (Goulder & Schein, 2013). The responsibilities, rights, and interests of all parties involved in the carbon market need to be clarified and laws and regulations must be enacted to provide guarantees for carbon trading. The government must take into account the strategic significance of reducing carbon emissions. There is a need to increase the content of the low-carbon economy in energy, environmental protection, and resource management laws and to promote the development of a national legal system to regulate the transformation process to achieve a low-carbon economy. However, as we will see, carbon credits are an emerging popular form of carbon pricing globally.

Carbon credits

As seen above, to date, carbon pricing can be done in two ways, through taxes on carbon and carbon-derived released into the atmosphere, which is carbon-intensive, or through carbon trading negotiations, the so-called cap-and-trade. In exercise, states, municipalities, and nations currently support some carbon pricing model.

The introduction of carbon credit was ratified in the Kyoto Protocol. The Paris Agreement validates the application of carbon credits to reduce greenhouse gas emissions and sets out the provisions to further facilitate carbon credit markets.

The main vehicles discussed for the creation of the low-carbon economy are carbon taxes and limit and trade schemes both depend on government action and both are driven in their public discussion by the issue of costs. Under this scenario, it can be argued that establishing carbon credits as an emerging alternative approach could complement regulatory schemes and work together to create a global carbon market at a pace that far exceeds current efforts. From this perspective, the role of the initiatives to be taken by private-sector financial institutions becomes crucial. Carbon credit and carbon finance are likely to be key to future economic decision-making and as this process unfolds.

At a greater level of understanding, a carbon credit is defined as a tradable license or certification that gives the holder the freedom to emit a total of one ton of carbon dioxide or an equivalent of any other greenhouse gas. The main purpose of this principle is the minimization of carbon dioxide emissions.

However, while this mechanism is attractive and, at present, an accessible strategy, since the regulatory agencies specify ceilings, some industries do not fit in immediately and therefore become economically unviable. In this sense, the possibility of acquiring additional carbon credits to affect the emission limit can reduce emissions immediately. Thus, segments that achieve carbon offsets, i.e., remove or avoid greenhouse gas emissions that were or would be emitted into the atmosphere and destroy, store or sequester them for periods of varying duration, depending on the sequestration method, are often rewarded with additional credits. Therefore, the sale of surplus credits can be used to subsidize future emission reduction projects, besides to choose a scheme that ensures permanent or long-term sequestration and creates local social and economic benefits.

Like carbon markets, there are two major divisions about carbon credits, namely voluntary emission reduction (VER), defined as carbon offsetting that is traded over-the-counter or voluntary for credits, while Certified Emission Reduction (CER) comprises emission units (or carbon credits) created through a regulatory framework to offset the emission of a project. The main difference between the two is that there is a third-party certification body that regulates certified emission reduction as opposed to voluntary emission reduction (Fig. 6.6).

Additionally, among the many environmental indicators developed over the years, some carbon credit quantification strategies have resulted in mathematical equations to strengthen this accounting. Through easy understanding, environmental indicators for quantifying carbon credits take into account direct

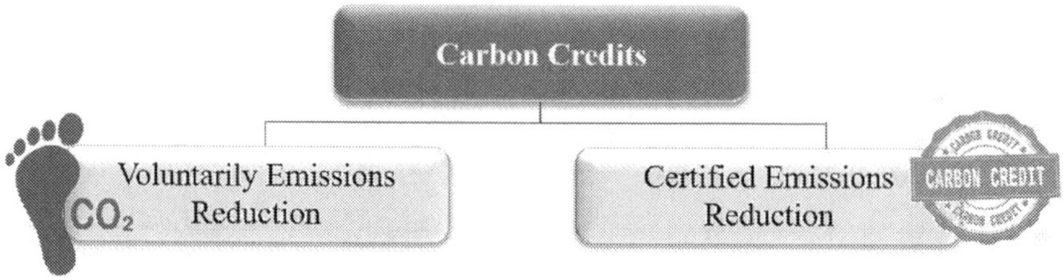

FIG. 6.6 Carbon credit category segments.
This figure illustrates the two main models of carbon certification.

CO_2 emissions (as a result of the process) and indirect CO_2 emissions (due to electricity consumption). Thus, these values are compared with direct, indirect, and therefore total plant emissions. Therefore, to quantify the potential benefits of a given process plant, two sustainability metrics are defined (Eqs. 6.9 and 6.10):

$$CO_{2NP} = CO_{2C} - CO_{2E} \tag{6.9}$$

$$CO_{2AV} = CO_{2NP} - CO_{2CP} \tag{6.10}$$

where CO_{2NP} represents the CO_2 not produced (kg CO_2eq); CO_{2C} represents the CO_2 conventional (kg CO_2eq); CO_{2E} represents the CO_2 emitted (kg CO_2eq); CO_{2AV} represents the CO_2 avoided in the process (kg CO_2eq); and CO_{2CP} represents CO_2 captured (kg CO_2eq).

Thus, while Eq. (6.9) takes into account the actual CO_2 savings, i.e., CO_2 not produced due to nonuse of fossil fuels as a raw material in the industrial process, Eq. (6.10) refers to the total amount of CO_2 that is not emitted into the atmosphere, i.e., CO_2 avoided, because (i) there is a certain amount of CO_2 that can be incorporated through process integration measures to be used as raw material, and (ii) some CO_2 is not produced compared to the conventional process.

Policy strategies among carbon footprint, carbon market, and carbon pricing

It is noteworthy that in the current context, the use of carbon footprint is intended to directly estimate and reflect on carbon pricing quantifications as well as their possible subsequent credits. As already seen, carbon prices take on several models. Usually, to circumvent commodity prices at the point of consumption, supporting carbon taxation at the point of production results in lower costs when related to the entire supply chain.

At the level of exemplification, the application of carbon prices reflects in the generation of market-based financing viability. That is, the price can provide a competitive advantage for companies with low-carbon profiles, thereby increasing the flow of financial resources to these sectors, supporting the transition to a low-carbon economy.

There are currently three internal methods companies use to assign carbon pricing, being carbon taxes, the shadow price of carbon, and implicit pricing carbon. Commonly, companies adopt a carbon tax, i.e., the monetary value per unit of greenhouse gas emissions, this cost is added to the operational expenses of the follow-up. However, by opting for such management, the company generates incentives to minimize or reduce short-term emissions while encouraging long-term innovation. Thus, an industrial follow-up can impose a carbon tax on all its business units or products, or just a few. Thus, carbon tax grades may be partially low to encourage the buy-in of individual business units.

Similarly, materializing all greenhouse gas emissions in monetary terms provides an effective alternative to promote energy efficiency through short-term emission reductions. Thus, the influence of each operating unit on an industrial follow-up through carbon taxation reflects in its decision-making and therefore generates benefits in a small atmosphere in order to modify the usual conduct. In addition, by implementing these changes there is the possibility of reducing costs by creating a property for climate mitigation. Charges may be used to increase carbon reduction targets that would otherwise require an alternative source of financing. This tool has the potential to define the company as a leader in climate strategy. This assumes that the company has the resources and experience to measure its carbon emissions in a standardized manner. The lack of a carbon reference price in areas without carbon pricing can be a barrier for a company to consider adopting this practice. Therefore, the internal price of carbon assumes to be a growing corporate practice.

The second method that industrial segments use to price carbon is known as the shadow pricing method. This method involves the company incorporating an internal carbon price in each of its investment plans in order to study the impact of greenhouse gas emissions on the company's return on each investment. In addition to its use in guiding capital investment projects, a company may also use a carbon shadow price as a proxy for future regulatory scenarios, as well as states of the world to model its impact on its business operations and resource mix. Thus, shadow carbon pricing can help a company in risk management as well as in strategic planning.

In a global context, in practice, several utilities are required by their regulators (utility commissions) to develop integrated resource plans (IRPs) to meet future energy demand and maximum capacity. The use of domestic carbon prices in IRPs has grown with a carbon credit.

The shadow price of carbon may be based on observed/anticipated carbon prices for established carbon trading schemes. Shadow prices can also be based on other government policies that implicitly price carbon, including the price of renewable energy or taxes on certain commodities. In addition, companies may also use a price range depending on a project's carbon exposure in a highly regulated market. Shadow prices used by companies tend to be higher than carbon tax prices.

Under implied prices, a carbon price is calculated based on how much the company spends on greenhouse gas emission reductions as well as regulatory compliance. In this case, the price of one ton of carbon is the marginal cost of reducing these emissions. This tool is backward compared to the first two carbon pricing methods described above, as it is based on the company's past decisions compared to what the company can do in the future. It is helpful to provide a company with a better understanding of its carbon footprint as well as the costs of regulatory compliance. Thus, a company may use more than one carbon pricing approach simultaneously and may also choose to apply a uniform price for all its operations or choose to follow a differentiated pricing strategy.

The carbon price range reported by the carbon disclosure project is quite large, from USD 1 to USD 800 per ton of CO_2 emissions. Usually, if there is an explicit price set by regulation, companies tend to use that price. Where there is mixed regulation or lack of regulation, prices tend to vary considerably. The main sectors that used an internal carbon price (84% and 79% of market value) in 2017 were the energy and utility sectors, respectively. Financial institutions, information technology companies, and consumer goods, although less carbon dependent, are also using this tool (24%, 27%, and 24% of market value, respectively). Based on the responses provided to the carbon disclosure project, high carbon sectors/companies used parallel pricing as opposed to a carbon tax. This choice of companies seems logical, as a carbon tax translates into an immediate cost to a company, as opposed to a parallel price. In addition, heavy carbon sectors face greater carbon regulatory risk.

In the current state, in countries where carbon pricing already works, the tax set at € 20 per ton of carbon generates € 440 million per year. However, future prospects assume that by 2030, carbon prices will reach at least € 80 per ton be subject to hypotheses—surrounded for a specific purpose—where funds are used to assist in the transition to a low-carbon economy, for the state project.

Given this economic perspective, two main state management strategies are proposed by the countries involved. The two paths outlined at the draft level require that any accumulated revenue be allocated to serve directly to the population living in extreme energy poverty, or to return profits through bonuses to all households.

However, preliminary trials reported through public reports indicated that the two main alternatives are not necessarily exclusive. Whereas, the first possibility aims to allocate any product as a compensation fund to those suffering from energy shortages in order to carry out energy reform by balancing greenhouse gas emissions while protecting climate change. Thus, the government project needs to introduce the consolidated carbon quantification associated with carbon tax conversion as well as carbon credit in order to reduce environmental impacts by increasing carbon taxes by projecting profits for society.

In contrast, the second possibility through carbon credit is in the form of fees and dividends. In short, this alternative does not differ so strongly from the former, however, incomes are returned to all citizens through a dividend to all households. Most households would be called net beneficiaries with those with lower incomes as potentially earning more. However, regardless of what is admitted in the medium and long term, there is much to be researched and consolidated before being put into practice by state projects.

Finally, regardless of the mechanism to be consolidated, only through a conscious planning methodology can society as a whole really consider the perplexing situation they are facing with respect to divergences between economic objectives and environmental protections (Felli, 2019). Moreover, it is only through this mechanism that they can systematize their concerns and change their beliefs, values, and demands through mutual interaction with other parties involved. The Climate Protection Agency would ensure that global limits, such as the limit on greenhouse gas emissions, are not experienced as external impositions but as part of the economic development to be created. Thus, this planning could help materialize and concretize the abstract issue of global warming, not just for its catastrophic consequences, but as an element of a broader process of production and democracy that could be judged not only by costs or gains economic. In view of these perspectives, decisions made by negotiations and deliberations would generally tend to be more robust and legitimate in the long range.

Carbon market trades

What is commonly called carbon market trading is a collection of various transactions by which greenhouse gas emission reduction volumes are negotiated and which distinguish between size, shape, and regulations. With the exception of these challenges, and while a seemingly insufficient emission reduction target had been agreed upon in the Kyoto Protocol, the timeframe since the Kyoto conference led to a series of institutional and market planning that potentially provided a basis for a postsettlement agreement. A relevant case to note is the decline in requirements for carbon credits, resulting from oversupply, uncertainties about the future of the carbon market, and the global economic crisis. Given this context, there has been a 17% elevation in the total volume of carbon credits traded in the last years. This expansion was affected by the increase in hedging and portfolio adjustments by participants, benefiting from the drop-in carbon credit prices.

Influenced by the postagreement, three organizations were structured to manage carbon markets. Thus, Joint Implementation (JI), Emissions Trading (ET), and the Clean Development Mechanism (CDM) were defined as the main carbon traders (Table 6.4). In sum, the first organizational structure of market traders aims to allow the implementation of carbon offsets and the implementation of projects considering carbon sinks in other Annex countries. The second structure makes it possible to delineate transactions associated with greenhouse gas emissions between the Annex countries. So, finally, the third organization gives industrialized countries the opportunity to fulfill their carbon

TABLE 6.4 Comparison of the different peculiarities of the main carbon markets.

European Union Emissions Trading Scheme	Clean Development Mechanism
European developed countries	Developing countries
Common in departments: Cements and lime, papers, irons, ceramics, chemicals, hospitals, universities, aviations	Common in departments: Energies, transports, fugitive emissions from fuels, solid fuels, oils and gases, industrial processes, mineral products, chemical industries, production of halocarbons and sulfur hexafluoride, consumers halocarbons and sulfur hexafluoride, agricultures, land uses, land-use changes and foresters, and sewage treatments
Emission limitations passed on to governments; designs change according to countries policies	Emission decrease targets are passed on to governments and project implementation is in accordance with countries policies
Target until 2020	Target until 2020
Annual limitations	Goals in the period and passed on to governments
Three phases	Two phases
Participation of the Europe Union is currently mandatory for certain industrial departments	Still voluntaries
Emission limitations (permission to emit, emission licenses, allowances)	Emission decrease targets (reduced emission reduction)
Greenhouse gases converted to CO_2eq	Greenhouse gases converted to CO_2eq
Annual monitoring	Annual monitoring
Reasoned on learning by doing initiate principle with updates and changes needed over time	Reasoned on "learning by doing" initiate principle with updates and changes needed over time
European countries participate in the programs	Interested developing countries can participates
Stages in the European Union Emissions Trading Scheme processes: Preparations, approvals, emissions permit transfers, monitoring, crediting, verifying	Stages in the CDM: Preparations, approvals, validations, registers, monitoring, certifying, carbon credits issuances
Taxation: national policies	Taxation: national policies

TABLE 6.4 Comparison of the different peculiarities of the main carbon markets—cont'd

European Union Emissions Trading Scheme	Clean Development Mechanism
Regulatory committee: Commission of Europe Union	Regulatory committee: Clean Development Mechanism Executive Commission
Methodology of calculation of emission decrease is more generalized	Methodology of calculation of emission decrease is more detailed
Legal nature of documents, arising from the DIRECTIVE 2003/87/CE and other amendments, are based on this initial document	Legal nature of documents, arising from the Kyoto Protocol of 1998 and other amendments, are based on this initial document

Based on Godoy and Saes (2015).

minimization responsibilities by investing in low emission projects in developing countries (UNFCCC, 2001).

The joint implementation comprises a bilateral and, in many cases, multilateral treaty, which countries with high pollution reduction or environmental conservation expenditures invest in the reduction or preservation of a country with the lowest expenditures. As a result, they acquire credits for reducing emissions or adding environmental protection. While joint implementation presents itself as a potential strategy for application in any environmental purpose, its efficiency can certainly be undermined as investing countries may face obstacles to achieving the maximum reduction in air emissions. In view of these application bottlenecks, it is necessary to define valuation projects that allow the quantification of indirect effects and market distortions through the integration of baselines associated with countries where the expected uncertainties are weighted. In this way, specific baseline scenarios are structured and passed on to the basis of joint implementation. For this reason, it is often applied in contexts where one partner of the agreement is committed to reducing pollutant emissions. In addition, this market has trade restrictions as emission reductions in the low-cost host country must at least offset the avoided reductions in the "donor" country. Therefore, the obvious potential attraction of joint implementation is that it reduces the overall costs of meeting internationally agreed emissions targets.

Emissions Trading Scheme (ETS), known as the first major emissions trading market, emerged in the mid-2000s, ruled by the United Kingdom. Years later, after the closure of the program, the European Union took command of the trading scheme and founded the Europe Union Emissions Trading Scheme (EU ETS). European leadership was primarily driven by early leadership to assist countries in achieving the targets set by the Kyoto Protocol. Thus, the creation of a Binding Directive acted as disciplinary conduct enabling the institutions comprised in the European Union Emissions Trading Scheme to utilize the carbon credits of clean development mechanisms to comply with their commitments (World Bank Group, 2014).

Nevertheless, as they could not depend directly on the clean development mechanisms, it was established that each participating European member would develop their own national allocation plan. These plans determined the amount of greenhouse gas emissions that could be reallocated to industrial and energy processes. Therefore, it was up to each manager responsible for their company to regulate its pollutant profile to remain within the established emissions quota. However, if the limit was exceeded, licenses could be purchased and, if below, sold. Thus, in the first moment, these organizations encompassed energy, metal, and cellulosic sectors, totaling about 12,000 industrial installations.

With increasing environmental concerns, some changes have been introduced to the programs. These include the integration of new industrial segments such as the aviation sectors, as well as the admission of new countries such as Norway, Iceland, and Liechtenstein. In addition, with the tightening of greenhouse gas management, other chemical compounds have been proposed. The capture of petrochemical and nitrogen volatiles, as well as carbon dioxide emissions related to transportation and geological storage, were inserted as prerequisites for meeting the targets. In addition, it is estimated that the reduction of the limits allowed by countries will increase substantially by 2020, as the number of allowances to be auctioned also tends to increase by 50% (Knox et al., 2012).

In these respects, the Europe Union Emissions Trading System has taken over more than half of the carbon dioxide emissions in the European Union. Moreover, more than just the global carbon market, its strategy has allowed the insertion of carbon allowances to developed countries. Thus, carbon has become a tradable capital throughout Europe. However, its low pricing has not made it possible to provide effective incentives to significantly reduce emissions.

Although the European carbon market is consolidated, other Clean Development Mechanism project credit markets are also well-established in the industrial community. Historically, the Clean Development Mechanism is an extension of the implementation of the Kyoto Protocol. Likewise, such as other carbon market negotiators, this organization aimed to introduce economic tools to assist parties that fulfill the objectives and principles defined by the United Nations Framework Convention on Climate Change. Thus, in order to ensure greater reliability on the intergovernmental issues associated with the climate change panel, the United Nations General Assembly has designated the Intergovernmental Negotiating Committee as the general planning director of the United Nations Framework Convention on Climate Change.

Once seen by many as a pioneering mechanism, the clean development mechanism has been known to enable a country with a commitment to reduce emissions or limit emissions under the Kyoto Protocol to implement an emission reduction project in countries on development. These projects build bridges with the relationship of obtaining certified emission reduction credits. In addition, it is well known that through it the first global organizational structure for investment and environmental credit was able to credit standardized emission compensation instruments.

One of the key differentials of the clean development mechanism, when compared to other structures, is its incentive to sustainably develop emissions reductions while allowing highly industrialized countries the flexibility to meet their targets. However, details of reductions should be clearly reported through qualified projects through a rigorous and public registration process. Consent is granted by the entitled national authorities. In addition, among other obstacles to be met, public funding for project activities should not result in the diversion of official development assistance. Therefore, the organization is overseen by the Executive Council, ultimately responsible for the countries that have ratified the Kyoto Protocol.

Given this global market context, while carbon-intensive sectors appear to be taking encouraging action, a regulatory response may be needed for less carbon-intensive sectors to leverage their market power to facilitate cooperative carbon reduction with broader constituencies. In addition, incentives for carbon-intensive sectors may be needed to encourage continued efforts to bridge the emerging carbon gap.

References and recommended reading

Papers of particular interest have been highlighted as:
• of special interest
•• of outstanding interest

BP. (2007). *What is a carbon footprint?* Accessed into site: http://www.bp.com/liveassets/bp_internet/globalbp/STAGING/global_assets/downloads/A/ABP_ADV_what_on_earth_is_a_carbon_footprint.pdf.

Carbon Trust. (2007). *Carbon footprint measurement methodology, version 1.1.* 27 February 2007 London: The Carbon Trust. Accessed on: https://www.researchgate.net/deref/http%3A%2F%2Fwww.carbontrust.co.uk%2F.

Dong, H., Geng, Y., Xi, F., & Fujita, T. (2013). Carbon footprint evaluation at industrial park level: A hybrid life cycle assessment approach. *Energy Policy*, *57*, 298–307.

Ellerman, A. D., & Buchner, B. K. (2007). The European Union emissions trading scheme: Origins, allocation, and early results. *Review of Environmental Economics and Policy*, *1*(1), 66–87.

Energetics. (2007). *The reality of carbon neutrality.* London www.energetics.com.au/file?node_id=21228.

ETAP. (2007). *The carbon trust helps UK businesses reduce their environmental impact.* Press Release http://ec.europa.eu/environment/etap/pdfs/jan07_ca rbon_trust_initiative.pdf.

Felli, R. (2019). Beyond the critique of carbon markets: The real utopia of a democratic climate protection agency. *Geoforum*, *98*, 236–243.

Finkbeiner, M. (2009). Carbon footprinting—Opportunities and threats. *The International Journal of Life Cycle Assessment*, *14*, 91–94.

Gao, T., Liu, Q., & Wang, J. (2013). A comparative study of carbon footprint and assessment standards. *International Journal of Low-Carbon Technologies*, *9*(3), 237–243.

GFN. (2007). *Ecological footprint glossary*. Oakland, CA: Global Footprint Network. Accessed July 2007 from http://www.footprintnetwork.org/gfn_sub.php?content=glossary.

Godoy, S. G. M. D., & Saes, M. S. M. (2015). Cap-and-trade and project-based framework: How do carbon markets work for greenhouse emissions reduction? *Ambiente & Sociedade*, *18*(1), 135–154.

Goulder, L. H., & Schein, A. R. (2013). Carbon taxes versus cap and trade: A critical review. *Climate Change Economics*, *4*(03), 1350010.

Grubb, & Ellis. (2007). *Meeting the carbon challenge: The role of commercial real estate owners*. Chicago: Users & Managers.

Hawkins, S. (2016). *Carbon market clubs under the Paris climate regime: Climate and trade policy considerations*. International Centre for Trade and Sustainable Development (ICTSD).

International Organization for Standardization. (2006). *Greenhouse gases—Part 1: Organization-level guidance specification for quantification and reporting of greenhouse gas emissions and removals*. ISO, 14064-14061.

Knox, J. W., Hurford, A., Hargreaves, L., & Wall, E. (2012). *Climate change risk assessment for the agriculture sector (DEFRA Project Code GA0204)*. London: Department for Environment. Food and Rural Affairs (DEFRA).

Patel, J. (2006). Green sky thinking. *Environment Business*, *122*, 32.

POST. (2006). *Carbon footprint of electricity generation*. POSTnote 268, October 2006 London: Parliamentary Office of Science and Technology. http://www.parliament.uk/documents/upload/postp n268.pdf.

Rodriguez-Lopez, J. M., Engels, A., & Knoll, L. (2017). Understanding carbon trading: Effects of delegating CO_2 responsibility on organizations' trading behaviour. *Climate Policy*, *17*(3), 346–360.

UNFCCC, & Conference of the Parties (COP). (2001). *Report of the Conference of the Parties on the second part of its seventh session, held at Marrakesh from 29 October to 10 November, addendum, part two: Action taken by the conference of parties*. United Nations Framework Convention on Climate Change. FCCC/CP/2001/13/Add. 1.

Wackernagel, M., & Rees, W. (1996). *Ecological footprint: Reducing human impact on the earth*. Gabriola Island, BC: New Society.

Wiedmann, T., & Minx, J. (2008). A definition of 'carbon footprint'. *Ecological Economics Research Trends*, *1*, 1–11. This chapter briefly shows logical reasoning for the creation of the term carbon footprint. It also brings several approaches regarding the different perspectives that gave rise to the theme. In addition, it presents numerous carbon footprint quantification proposals through online calculators.

Wintergreen, J., & Delaney, T. (2007). ISO 14064, international standard for GHG emissions inventories and verification. In *16th annual international emissions inventory conference, Raleigh, NC*.

World Bank Group. (2014). *Doing business 2015: Going beyond efficiency: Comparing business regulations for domestic firms in 189 economies: A World Bank Group flagship report*. World Bank Publications.

Chapter 7

The role of life cycle assessment in agricultural systems

Agriculture is widely regarded as one of the most vitally important sectors for global economic expansion. However, this sector assumes a crucial role in food security, and at the same time, it becomes responsible for significant environmental issues. As a result of the growing world population, the commitment for agriculture to increase food production became evident, while, consequently, inflicts increased environmental impacts (Tsangas, Gavriel, Doula, Xeni, & Zorpas, 2020).

Within this scenario, agriculture faces unprecedented challenges in finding a balance between increasing productivity and environmental protection, in addition to dealing with the rapid growth and increase in urbanization rates of the world population.

According to recently published estimates, contemporary agriculture is the main activity responsible for accelerating the loss of biodiversity. In addition, it is among the main practices responsible for increasing factors of climate change and human-induced changes to the nitrogen cycle. For this reason, interest in designing sustainable agricultural systems as an integral part of sustainable economics becomes relevant.

Under global aspects, it is imperative for the agricultural sector to take responsibility for adapting to the tangible need to develop an environmentally rectified food production, taking as a reference mitigated environmental impacts to ecosystems, as well as resulting in decreases in carbon dioxide emissions. And, therefore, reduce the overall environmental burden.

Known for being a sophisticated environmental management tool, life cycle assessment (LCA) is an alternative device to robustly assess the main environmental impacts of agricultural systems, as well as industrial systems. In addition, it is generally applied in these areas, since it serves as a reference for the construction and implementation of assertive policies, business definition, as well as a key tool for decision-making. Thus, it is suggested that the practice of life cycle assessment has been well documented as a tool to compare products and processes or to compare various components in a life cycle (Brentrup, Küsters, Kuhlmann, & Lammel, 2004).

Sustainability Metrics and Indicators of Environmental Impact. https://doi.org/10.1016/B978-0-12-823411-2.00007-4

Despite presenting a broad spectrum of application and development, and receiving considerable interest from governments and the scientific community, the LCA tool lacks an overview of research in the field of agricultural and food economics. This is because, in fact, we are used to applying the LCA methodology to quantify and evaluate the demand for natural resources and the possible impacts emitted by processes and products. Therefore, the production system must be considered. However, in LCA assessments in agricultural production systems, the scope of the evaluation should include not only activities in the field but also all other stages associated with the production of raw materials. That is the consumption of minerals and fossil fuels, as well as requirements regarding agricultural inputs, such as fertilizers, substances, machinery or plant protection seeds (Audsley et al., 1997).

Based on the bottlenecks reported above, the following topics aim to facilitate the application of the LCA for agricultural systems to obtain more accurate results from environmental sustainability indicators. Besides, we will clarify some essential deficiencies regarding the lack of integration of associated impacts for agriculture, such as inconsistencies in the definition of the functional unit, issues related to land use, consumption, and accounting of agricultural inputs.

Functional unit for agricultural systems

The functional unit approach to agro-industrial systems requires a broader perspective of product-environment interactions at all stages of the cycle. In short, in all assessment systems, the definition of the functional unit can lead to high levels of uncertainty and result in the following limitations of process comparison.

Evidently, the choice of the functional unit of agricultural systems needs to present representative functional units. Besides, they must have the usual effectiveness regarding the data generated by environmental impacts. Therefore, it is expected that these units should be carefully selected to develop the ideal sample design for impact measurements and to ensure the correct mapping of environmental burdens throughout the process.

In general, the main function of an agricultural system is to produce an item for human or animal use (food or nonfood), which may require additional processing (cooking, cooling, or mixing) before it can be used. Thus, we commonly encounter agricultural LCAs, whose functional units are defined as the basis for calculating 1 kg of product. In terms of exemplification, it may seem logical to establish the functional unit of 1 kg of wheat flour as the primary raw material for bread production. However, the definition of the scenario may be more complicated than it appears, since 1 kg of wheat flour does not necessarily present the same nutritional conditions as other wheat flours produced in different locations. What we should pay attention to is in line with the product's most intrinsic characteristics. That is, the variation may occur due to inequalities in the levels and types of proteins and other nutrients, as well as mainly the different enzymatic activities attributed to the quality of the wheat grain. This variation is determined by several factors, including the variety of wheat, production method, climate, and conditions of drying and storage. Therefore, for LCA assessments in agricultural systems, depending on the degree of detail of the analysis, it is important to recognize the patterns of food consumption, which must be incorporated into the LCA structure when they identified factors that limit sustainability, such as the rapid conversion of primary agricultural land, excessive depletion of topsoil, and illegal agricultural operators in the region.

At the same time, one of the crucial characteristics in the field of LCA applications in agriculture is the use of several functional units. In fact, it is quite common to come across different functional units. For example, in the agricultural sector, we usually find analyzes that use functional units per farm unit for a specific plantation area, for cattle units or the final product. Thus, the plurality between the functional units found in agricultural systems can generate a multitude of discussions about the multifunctionality of agriculture and, consequently, result in different ways of interpreting the results.

These units recently launched technical debates about the multifunctionality of agriculture. That is, the environmental impacts are regularly expressed by units of product mass per unit of applicable land area (kg/ha). When carrying out the interpretation of the charges, it is possible to point out the essential function of production concerning the occupation of the field. However, it is strongly recommended that the functional units should include ecological conditions, socioeconomic parameters, as well as human interactions, from their interference in production to the use and deposition of the final product. Thus, the economic activity generated, that is, the gross profit is also recommended as a functional unit to express the financial situation of the analysis.

Given this perspective, it is possible that the use of two functional units at the same time can be suggested to enhance and discuss in a more assertive way the differences contained in the study. Additionally, an amplification of the limits of the system is also encouraged since it can be imperative for analyzes, especially when there is the use of renewable raw materials.

System boundaries

As seen in Chapter 2, the system boundaries are responsible for differentiating the system under analysis from its external environment and delimiting the study. However, it is known that many aspects can be taken into account for the establishment of acceptable limits to the system under analysis. The system boundaries can be presented as basic models subdivided into infrastructure and operation and must be supplied together with the functional units provided (Fig. 7.1). In short, the system boundaries must contain the necessary inputs and outputs for the operations of the activities included in the study.

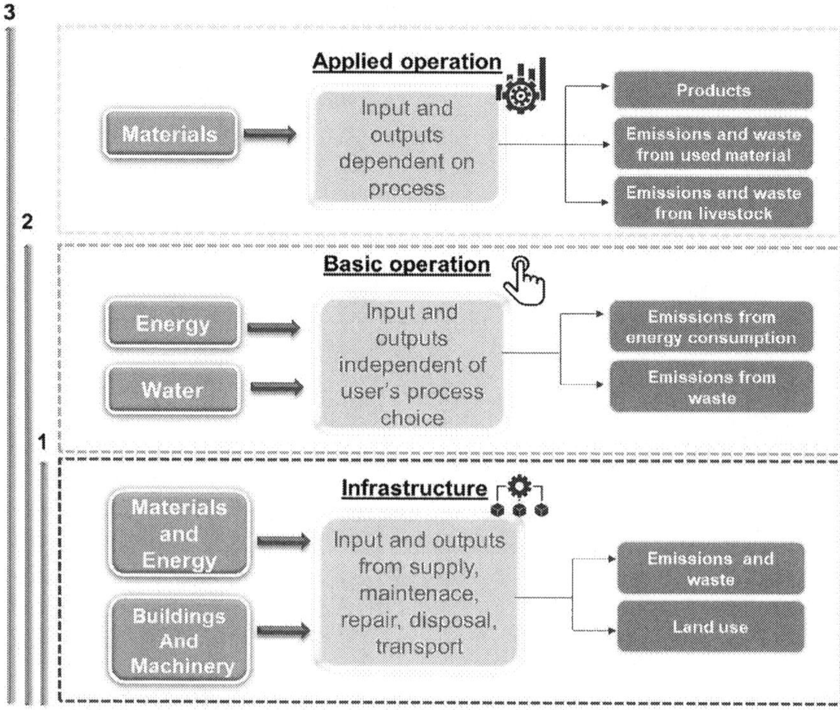

FIG. 7.1 The base model of infrastructure and operation associated with system boundaries. This figure illustrates a structural basis of evaluation parameters included within different levels of evaluation of the limits of the system applied to agricultural systems.

As can be seen in Fig. 7.1, the first structural class of the model is related to infrastructure. It is known that the infrastructure modules (buildings and machines) include, respectively, the construction and production of the infrastructure unit, as well as transport (raw materials, engines), maintenance, repair, and final disposal. On the other hand, when we move to the second level of assessment, it consists of considering the basic operations included in the agricultural system. In short, the basic operation modules comprise inputs and outputs independent of the user's choices, as well as the infrastructure used (level 1). The fertilization by emitter module, for example, includes the amount of support used (tractor, fertilizer distributor, and shed), as well as diesel input and emissions related to the use of the tractor (exhaust gases, tire wear) as the module can be used for a variety of different fertilizers (Brentrup, Küsters, Lammel, Barraclough, & Kuhlmann, 2004). This way, the fertilizer itself and all emissions related to the spread of the fertilizer in the field (for example, nitrogen and emissions containing phosphorus, heavy metals contained in the fertilizers) are not included in this assessment. Finally, the third and most complete class of evaluation consists of the systemic evaluation of the applied operations. The applied operations cover all inputs and outputs related to the operation that is dependent on the user's choices, as well as on the infrastructure used (level 1) and in basic operation (level 2). Examples of such inputs are fertilizers, seeds, pesticides, corral manure, animal feed, etc. Besides, field emissions of fertilizers and pesticides, livestock emissions, and emissions from the storage of manure from the corral can be considered as some examples. However, regardless of the specific level of assessment chosen, it is essential to note that from this choice, it will be possible to name the analysis, whether it will be from the cradle to the tomb or its variations.

Inherent in these issues, the processes for defining system boundaries must pay attention to the validity of the data obtained from the definition of the system boundaries. This is because this factor is one of the most pertinent concerns in life cycle analyzes. Due to globalization and the fast pace of research and development, new materials and manufacturing methods are continually being introduced to the market. Not unlike, in agricultural LCAs, this obstacle remains. However, does this question arise from the doubt between determining the limit of the cradle to the farm versus the cradle to the grave?

Obviously, it became evident that avoiding the complications of a more complete LCA study by defining a restricted system boundary that covers "cradle to the farm" instead of "cradle to the grass" causes more complications in the analysis than would be avoided in one study most complete of the LCA. Two examples are the definition of the functional unit (discussed above) and the assessment of nutrient depletion. Nevertheless, it is commonly considered appropriate to truncate the LCA on the farm for projects, due to its focus on agricultural systems. Note that the impact assessment still considers the impacts from the cradle to the grave (Fig. 7.2).

In addition, the system limits suggested in the LCA assessments indicate the use of agricultural limits and time. An agricultural limit is understood as one whose analysis aims to compare identical products when they pass through the farm gate. However, the products passing through the farm gate are not identical. And, therefore, it may be necessary to include other industrial processes, for example, drying outside the farm.

On the other hand, the time limit requires that the data collected should be selected based on the expected values for an average technology for a representative type of farm chosen as a typical system in a country. For this case study model, the opinion of experts and the literature are usually consulted, where there is more accurate information regarding the time to harvest.

Another parameter that is commonly overlooked or accounted for separately in agricultural LCAs is associated with capital equipment. The construction and maintenance of agricultural machinery should be included since some estimates from studies referring to evaluation models suggest that this parameter could contribute significantly when added to the other results of the impacts. In addition, the service life and utilization rate of the machines may differ from source to source. Therefore, correlating the underlying assumptions and the samples from which the data were derived should be considered in an attempt to explain these differences. Consequently, it can be concluded that it is possible to choose the data that most closely matches the actual data on environmental burdens.

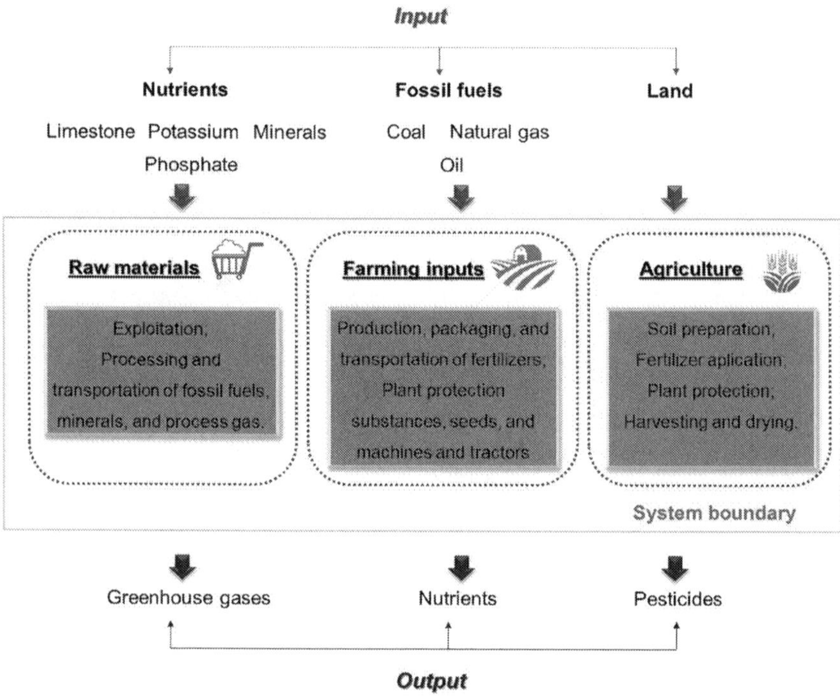

FIG. 7.2 A characteristic example of systems boundaries in the application of LCA in agricultural systems.
This figure shows some examples of inputs and activities that can be included in the boundary of the evaluated agricultural process system. In addition, it emphasizes, through the arrows, the entrances and exits of the system.

Likewise, crop rotation, that is, a system of sequential planting of plant species over the same area, so that this sequence is repeated after a certain period, is strongly recommended when inserting LCA analyzes. In fact, one of the central axes of sustainable models of plant production is the diversification of activities in the areas of cultivation to increase the ecological, economic, and social sustainability of agriculture. However, aware of the importance of this activity, few studies found in the literature are concerned with inserting this parameter. This is because, although a particular culture is considered to be the central focus of the studies, this, in turn, will interact with other cultures in a crop rotation. Therefore, these interactions must be included in the LCA of any crop, possibly through the evaluation of a soil quality index.

Allocation

As widespread, the allocation is known as the direct consequence of the multifunctionality of a product system. In addition, it is responsible for determining each product in a multifunctional system on environmental impacts. Whenever possible, the allocation should be avoided due to the high degree of uncertainty in the allocation of responsibilities.

However, as in other systems, agriculture allocation can also be avoided by expanding the operations (Fig. 7.3A). On the other hand, when it cannot be avoided, the allocation must be converted into different units, that is, reallocated in mass, energy, or price. Consequently, this allocation conversion may result in the generation of uncertainties regarding the results obtained.

However, the allocation of environmental impacts to coproducts based on their economic value is the most widely used allocation method in agricultural LCA studies, particularly in agricultural production and the animal supply chain. This choice refers to a more simplified procedure, the allocation of which is to price each product relatively at the point in the supply chain since the products will be separated. The central concept behind this economical product allocation model is that the current cause for production, and employing environmental interventions, is attributed to the economic value of the products (Mackenzie, Leinonen, & Kyriazakis, 2017).

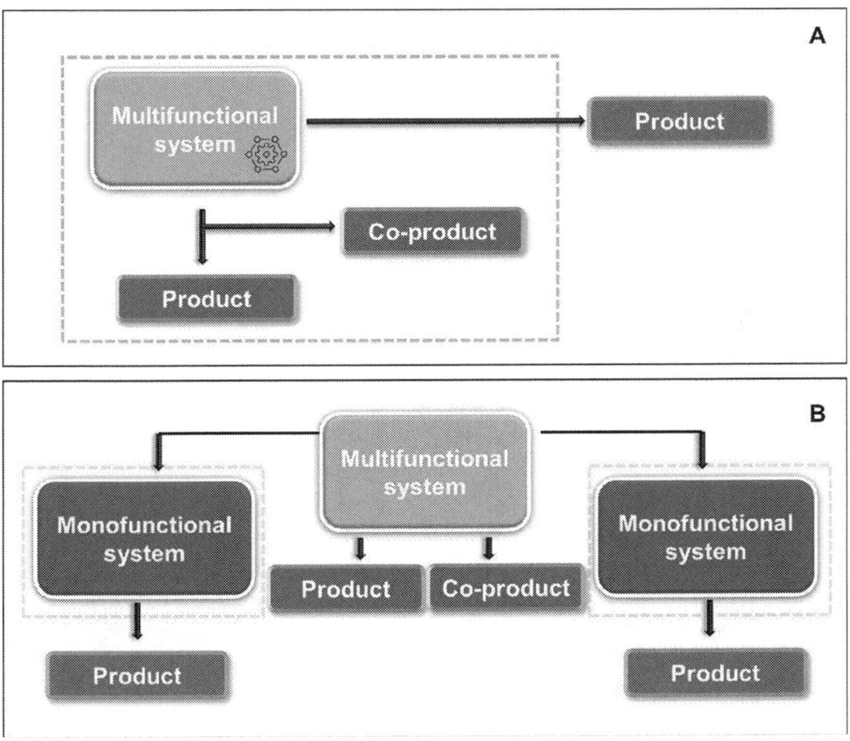

FIG. 7.3 Allocation models and ways to avoid double counting of environmental impacts. This figure generically shows two evaluation models to avoid possible allocations in the system under evaluation.

On the other hand, the allocation of environmental charges to different crops in one rotation may be relevant for various inputs and products during the specific period under analysis (Pawelzik et al., 2013). At the level of exemplification, commonly in the fertilization stage of cropping systems, the use of nutrients from manure and chemical fertilizers, the nutrients may not all be absorbed by the crop in the system under study. Still, they will likely remain available for future crops in the rotation. As a consequence, the environmental charges associated with the use of manure or chemical fertilizers must be allocated to each crop in rotation, depending on the nutrients required by each crop.

It is known, however, as a general rule, emissions associated with nitrogen, phosphorus, and potassium should be allocated in two ways: (i) the recommended amount for each crop, if considered alone without knowledge of the previous crop, or (ii) crop uptake, more emissions to air and water, due to crop growth.

Usually, in agricultural processing, a given harvest requires higher nutrient demands to grow successfully than those removed in the harvest. As a result, the latter option is not advisable because it does not explain the capture efficiency. Thus, where there is the possibility of combining different nutrients, for example, in manure, the allocation can be made by (a) the most limiting nutrient or (b) according to the economic value of the different nutrients (Bergström et al., 2005). In the latter case, the environmental charges associated with the use of manure are allocated to the system under analysis in proportion to the financial value of the nutrients used by the system compared to the total nutrients in the manure (allocation by economic value).

In contrast, another simple procedure can be applied for the different uses of organic matter considered as a nutrient to increase soil fertility in a broad sense. As all cultures benefit, the allocation must be based on (area $\times$ time) per culture. However, when there is no data available on the crops assessed in question, it is therefore suggested that all environmental charges associated with the use of manure and chemical fertilizers be allocated to the final product. However, another usual possibility is to consider the surplus of nutrients and any increases in organic matter in the quantification in categories of change and land use or also in indices referring to soil quality.

Simultaneously, for use on machines and buildings, the allocation is performed according to hours of use compared to total hours of use over the life of machines (for example, 12,000 hours). The main area where this is questionable is for tractors used for many different tasks, imposing very different levels of wear on the tractor. Engine speeds or kWh compared to the total over the life of the machine may provide better bases for allocation, but this approach would require more detailed data than that are available. Therefore, in LCA studies, the effect of the use on the machine's service life is generally disregarded (Fig. 7.4).

In addition, it is important to emphasize that the working time requirements for a given activity are influenced by the size and length of the plot and the width of the machine. The calculation of working time includes not only the working time of the activity in the field but also the setting time and the time

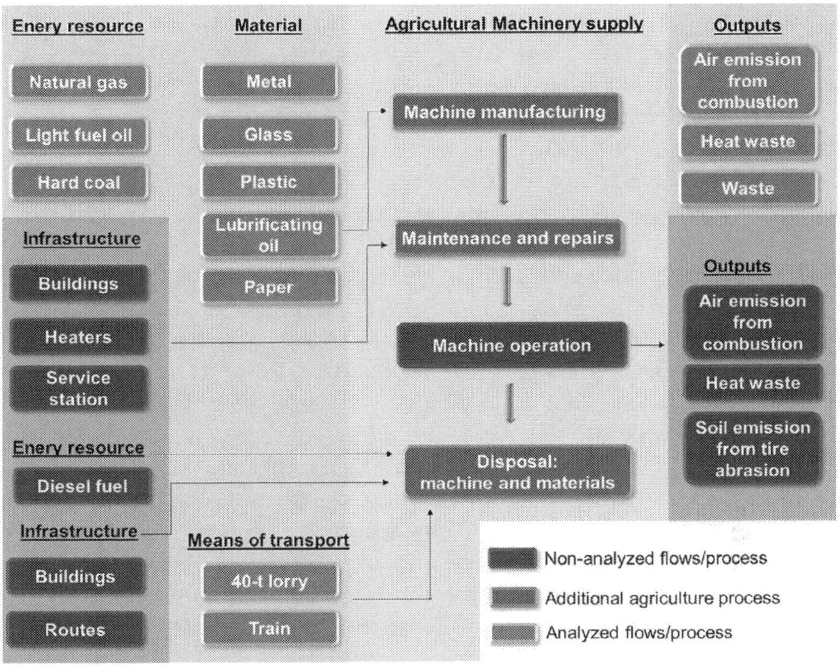

FIG. 7.4 Example of a flowchart of the system boundaries where important steps such as machinery and infrastructure are disregarded.
This figure presents an agricultural system model under life cycle assessment. However, through the markings, it is possible to note that not all stages of the process are necessarily considered in the evaluation. Therefore, it is up to the analyst to determine what will or will not be considered for the LCA and make it very clear what were the boundaries of the evaluated system.

to cover the distance between the farm and the field and for operations linked to the activity.

Quantification bottlenecks in the life cycle inventory stage of agricultural systems

A vital premise regarding agricultural products is the established argument that a significant part of farm products is grown in regions or countries other than those that are actually consumed. It is estimated that the central studies involving agricultural LCA, at the moment, disregard the stages of commercialization and costs of transformation and use of land, in addition to the main chemicals and fertilizers available to the environment. Thus, it is necessary to establish easy-to-apply and straightforward indications to standardize the inventory quantification stage of the life cycle analysis of agricultural systems.

Evidently, agricultural LCA approaches reflect more significant difficulties of the analytical establishment when compared to industrial systems, since this tool was first developed exclusively for industrial application. Later, recent advances were proposed to evaluate agricultural systems, and, therefore, the complexity of adapting the tool to the differences between these systems is notorious.

To circumvent these barriers, Mourad, Coltro, Oliveira, Kletecke, and Baddini (2007) developed a specific questionnaire for agricultural LCAs, which serves as a guiding principle for data collection for inventory analysis (Fig. 7.5). For this, it is essential that the filling of these questions is carried out formally by professionals in the area or by technicians responsible for the production system to facilitate the obtaining and reach of reliable data.

In addition, the questionnaire must cover as many details as possible. Thus, the document needs to include process specifications as all inputs used, that is, the insertion of nutrients, fertilizers, agricultural correctives, the use of pesticides, water demand, energy sources, whether for the use of transport and machinery diesel, as well as the electrical demand for the other equipment used. Another critical detail is the concrete establishment of the time of use for each equipment or activity performed. Information on product distribution to the export port should also be requested, such as the model and function of the truck used, as well as the total distance traveled. On the other hand, aspects of the crop such as general data on the farm, total area, area cultivated, plant density, cultivated varieties, types of harvest and harvest management, number of employees should also be collected (Fig. 7.6).

Given the above, even if we follow a certain parameter, within the environmental analyzes it is accepted that there are several different ways of evaluating the elapsed charges. In addition, another difficulty in implementing this tool in agricultural systems is the versatility of propagating the same cultivar, even when located in a given region, or under the same climatic conditions, as well as under similar soil types. Therefore, the choice of comprehensive and representative parameters in view of data consistency should be encouraged.

Thus, planning and building a spreadsheet with consistent data means that these systems are capable of reliably illustrating the choice of scope under study. These data are often based on procedures similar to previous studies. For this reason, the use of internationally recognized databases is necessary and is strongly recommended to serve as an analysis model.

On the other hand, even if all the care in selecting the data collection has been adopted, it is extremely difficult to maintain the identical reproduction of the systems under evaluation. This is highly attributed, given that the analysis almost always deals with estimated theoretical data. And, therefore, few studies report, in fact, real data from the systems. Thus, a fundamental strategy is based on the subsequent discussion of the data obtained when comparing them with the previously reported literature. Consequently, we must be aware that the differences between the data reflect real differences between the systems and not the variations in the data collection.

Parameters	Unit	Average/(kg/ton.)	Average (kg/ha)
Input			
Energy			
Total	MJ		
For cultivation	MJ		
For processing	MJ		
For transport of the chemicals	MJ		
Other Resources			
Water for product processing	kg/m³		
Fertilizers			
Total (actives and fillers)	kg		
N, P, K (macronutrients)	kg		
B, Cu, Fe, Mn, S, Zn (micronutrientes)	kg		
Pesticides	kg		
Total (actives and fillers)	kg		
Fungicide	kg		
Herbicide	kg		
Bactericide	kg		
Acaricide	kg		
Inseticide	kg		
Correctives			
Total (actives and fillers)	kg		
Ca, Mg	kg		
Land Use			
Land Use	ha/yr		
Output			
Organic residue used as fertilizer	kg		
Wastewater	kg/m³		

FIG. 7.5 The base questionnaire adapted to circumvent LCI bottlenecks in agricultural systems. *(Based on Mourad et al., 2007.)*

Parameters	Unit	Average
Productive plants	Plants	
Density	Plants/ha	
Yield	ton/ha	
Pluviometric index	Mm/yr	
Solar radiation index	$MJ/m^2/yr$	
Labour	labours hours/ton	
Plant varieties		
Soil cassification		
Crop rotation	Comments about tillage sequences, period, etc.	
Main agricultural management Practices	Comments about seeding type, soil preparation, soil covering, irrigation system, manual or mechanical harvesting, characteristics of the processing, etc.	

FIG. 7.6 Information example to be compiled about the farm data. *(Based on Mourad et al., 2007.)*

Land use

Unquestionably, agriculture is very intensive in terms of land use. The issue of land use and transformation in agricultural systems, by far, is among the most important issues in life cycle analyzes. The argument regarding the environmental impacts related to this activity is quite broad. It has been increasingly disseminated in depth by several researchers, whose measurement of human influence on ecosystems and guidelines for the application of the concepts of land use, change of use from the earth.

Some concepts that make up this category arouse interest in characterizing the main differences between the attributions to environmental charges related to changes in land use. It is known that these charges are distinguished between immediate causes of change in land use and underlying driving forces, where close causes are human activities or immediate actions that occur in a location. In contrast, the underlying driving forces denote the fundamental social processes that lead to causes. However, changes in agricultural land are generally identified as close agents; that is, the cultivation of new fields can cause the conversion of wetlands or also deforestation in tropical forests. In this sense, the conceptualization of nearby causes and underlying factors requires adjustment to study changes in the use of agricultural land. Instead of addressing nearby causes, the difference in the use of agricultural land can be characterized by the manifestations of these changes (Scherer et al., 2020).

At the same time, it is known that the most significant direct factor in the loss of terrestrial and freshwater biodiversity is land use, mainly due to agriculture. However, despite the reduction efforts, no improvement in biodiversity decline has yet been achieved. On the other hand, in most of the world's land surface, the loss of biodiversity has already exceeded the proposed planetary limits. As a result, this highlights the urgency to safeguard biodiversity and make better-informed decisions based on scientific evidence.

Biodiversity in agricultural landscapes depends mainly on the intensity of land use. Thus, a given assessment of changes in agricultural biodiversity needs spatially explicit information on the intensity of land use. Therefore, parameters are required to be able to distinguish between input intensity, which is measured by input variables, for example, chemical fertilizers, pesticides, and output intensity, measured as production per unit of land area and time.

Several quantification alternatives have been proposed to measure the loss of biodiversity. As a model characterization, we can assume the trend indicators in the abundance and distribution of selected species. This model is widely adopted by the European Union, where it appears on the list endorsed by the stakeholder conferences.

Conceptually, biodiversity is expressed as the average abundance of species originally present in natural ecosystems in relation to their abundance in undisturbed situations, which are called ecosystem quality. The establishment of the scale consists of defining a maximum value; that is, the percentage reference of 100% would indicate an undisturbed natural situation. In comparison, the minimum value (0%) represents a wholly modified or damaged ecosystem without any wild species. In short, this indicator is similar to the concept of the natural capital index (NCI). However, NCI is defined as the product of the size of the remaining ecosystem (quantity) and its quality (Eq. 7.1). In addition, NCI can be established for natural areas such as forests, inland waters, and pastures, as well as in artificial areas, such as agricultural land and urban areas.

$$NCI = Q_T \times Q_L \tag{7.1}$$

where NCI is the natural capital index (%), Q_T is the quantity of the ecosystem to be analyzed (%), and Q_L is the indicator of ecosystem quality (%).

As can be seen generically in Eq. 7.1, the amount of ecosystem is calculated as a percentage of the total area (% of the area of the country or region). The quality of the ecosystem is calculated by counting the average abundance of the main set of characteristic animal and plant species. Furthermore, conceptually, quality can be determined as the ratio between the current situation and the baseline state (percentage of the baseline).

At the level of exemplification, Fig. 7.7 presents the model diagrams of the quantity and quality of the ecosystem. As can be seen, the process of ecosystem degradation can be visualized using natural index capital. If, for example, 50% of the natural area remains with a quality of 50%, the NCI is 25%. This means that the average abundance of the original species is 50% of nature or low impact state and so on. To avoid masking, a significant increase in the

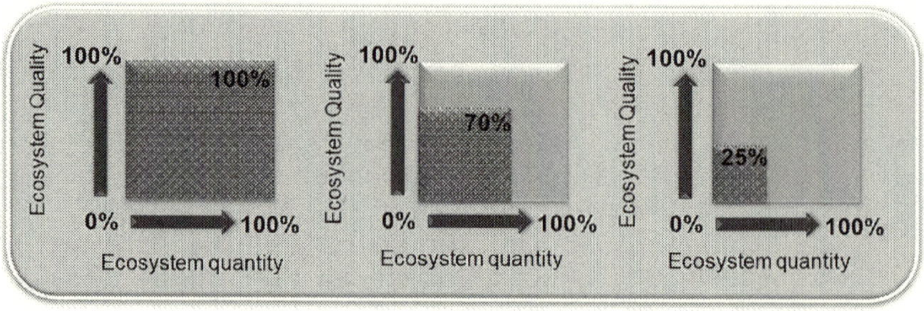

FIG. 7.7 Variation model of the natural capital index.
This figure illustrates the ratio between quality and quantity deteriorated of the ecosystem.

population of original species is truncated to 100%, although they must actually score negative. Thus, exotic or invasive species are not part of the indicator, but their impact is represented by the decrease in the abundance of the original species that they replace. Therefore, the NCI ranges from 0% to 100%, representing a deteriorated whole and an intact ecosystem, respectively.

Thus, to determine the dose-effect relationship of the intensity of agricultural land use in biodiversity, gradients of loss of biodiversity in land use must be quantified to make a comparison between land use systems and the primitive or low human impact. Still, it is suggested that a comparative peer analysis of the implications for biodiversity be applied between the different types of farms, recommending that they are specific to cultivation and pasture systems. When possible, a peer comparison of the impact on biodiversity for converting conventional agriculture to organic agriculture is also strongly recommended. For this purpose, Table 7.1 presents a precursor study of the theme, carried out by Reidsma et al. (2006), where the researcher provides indicators of ecosystem quality, focused on the combined implications of agricultural operations (frequency of mining, applications of fertilizers and pesticides and specialization of production, monoculture and crop productivity).

As can be seen in the previous study model, land use change (LUC) has often been considered harmful, referring to the impacts of deforestation and the expansion of biomass plantations. Currently, many ecosystems are being degraded or used in an unsustainable manner, compromising their ability to support multiple systemic services over time (Venter, Scott, Desmet, & Hoffman, 2020). In fact, through the ecosystem quality indicators, the impacts of land use as an input on the quality of the ecosystem or as a protected area can be quantified.

There are several LCIA methods for characterizing the impacts of land use on the quality of the ecosystem. They include, for example, net primary production, hemerobia naturalness, and, more often, species richness as indicators of ecosystem quality. While species richness is just a count of species, functional diversity considers the functional traits of a species. It is more strongly related to the functioning of the ecosystem than species richness.

TABLE 7.1 Synthesis of ecosystem quality models applied to different agriculture land profiles.

Ecosystem quality (%)	Farm models	Production systems
100	No production	Primary vegetation
40	Extensive grassland management	Medium to high cattle density on natural grassland
35	Extensive organic farm	Low-External-Input and Sustainable Agriculture (LEISA), permaculture
25	Extensive farming	Traditional farming; extensive farming; Low-External-Input Agriculture (LEIA)
20	Intensive organic farming	Rainfed organic farming
20	Intensive grassland management	Grassland production based on plowing, reseeding, and fertilization
15	Highly intensive organic farming	Organic farming in developed countries (where conventional agriculture is based on long-term soil and water investments)
10	Intensive production systems	Intensive agriculture; integrated agriculture; High-External-Input Agriculture (HEIA); conventional agriculture
5	Highly intensive production systems	Irrigation based agriculture; integrated agriculture; drainage-based agriculture; additional soil leveling practices; regional specialization; specialization of production at the farm and landscape level

Based on Reidsma, Tekelenburg, Van den Berg, and Alkemade (2006).

In addition, another way of assessing soil quality or how each impact can be attributed to insufficient or degraded ecosystem service provision under current agricultural practices can be viewed according to Table 7.2—the relationship between environmental impacts and the spatial indicator used for impact classification. According to the above, each landscape was classified as very low, low, medium, high, or very high: (i) nutrient emissions to water, (ii) loss of soil by water erosion, (iii) loss of soil by erosion wind power, (iv) recurrent flooding, and (v) accumulated loss of soil organic carbon (SOC). This classification was made using spatial indicators, according to Englund et al. (2020).

As Table 7.2 shows, soil loss from erosion was indicated by average annual soil loss from water erosion. Among this indicator, the practices of rain

TABLE 7.2 Spatial indicators, units, and thresholds for classifying landscapes as having varying degrees of negative environmental impacts, and the corresponding ecosystem service that is required for impact mitigation.

Environmental impact	Ecosystem service required for mitigation	Spatial indicator	Unit	Degree of environmental impact				
				Very low	Low	Medium	High	Very high
Soil loss by water erosion	Mass flow regulation	Annual average soil loss by water erosion on land used for production of annual crops	t soil loss/ha/y (Average value on land used for production of annual crops)	0	0–2	2–5	5–10	>10
Soil loss by wind erosion	Mass flow regulation	Annual average soil loss by wind erosion on land used for production of annual crops	t soil loss/ha/y (Average value on land used for production of annual crops)	0	0–2	2–5	5–10	>10
Recurring floods	Water flow regulation	Share of landscape area subject to 10-year flooding	% landscape area	0	0–5	5–10	10–25	>25
Accumulated loss of soil organic carbon (SOC)	Soil organic matter formation and composition	Soil organic matter formation and composition	Ratio of current SOC divided by theoretical max SOC (Average value on land used for production of annual crops)	>0.844 +null	0.688–0.844	0.532–0.688	0.376–0.532	≤0.3384
Accumulated loss of SOC – low estimate (−10%)				>0.759 +null	0.619–0.759	0.478–0.619	0.338–0.478	≤0.338
Accumulated loss of SOC – high estimate (+10%)				>0.928 +null	0.756–0.928	0.585–0.756	0.413–0.582	≤0.413

Based on Englund et al. (2020).

erosivity, soil erodibility, cover management, topography, and support were modeled with the most recently available European data sets. Likewise, the degree of soil loss due to wind erosion was estimated and classified as described above for water erosion, however, combining soil properties and daily data on rain, wind speed, evapotranspiration, soil moisture, and land cover harvest canopy. On the other hand, to indicate the degree to which individual landscapes are prone to recurrent flooding, the share of the total area in each landscape subject to 10-year flood events for each landscape was calculated. The data were obtained using a cascade model simulation approach composed of the configuration and calibration steps of the distributed hydrological model, simulation of long-term discharge time series and derivation of peak flows with selected return period, scaling down to the spatial resolution of 100 m and derivation of project flood hydrographs, and hydraulic floodplain simulations and fusion of output flood depth maps. In contrast, the accumulated organic carbon losses from the soil were calculated using the SOC's average saturation capacity on land used for annual crop production. As a consequence, the limits were defined using geostatistical properties to determine five equal intervals between the minimum and maximum aggregate mean values of the SOC's saturation capacity, based on recommendations from experts (for example, model developer). In addition, for these impacts, high (10% higher limit values) and low (lower 10% limit values) estimates have also been defined to allow for a sensitivity test.

Fertilization of agricultural crops

The deposition of nutrients and the use of pesticides has become an essential activity in modern agriculture. Efforts for crops to show high growth rates and yields have made this practice an attractive alternative to support the demands for agricultural products. However, despite its apparent benefits, it is known that its excessive application has numerous negative consequences for the ecosystem.

Nitrogen sources are considered one of the main nutrients used in agricultural applications. Its most diverse forms can be found through its nonreactive form (N_2) or through the air, soil, or water in any of its reactive forms, such as a nitrous oxide (N_2O), nitrate (NO_3^-), or ammonia (NH_3). Consequently, nitrogenous substances are recognized for playing a significant role in plant growth and development, causing atrophy when absent. However, when nitrogen in its reactive form is released into the environment, its environmental impacts can be very harmful to the ecosystem.

Thus, the quantification and location of the occurrence of the environmental impact are strongly indicated to be accounted for in the life cycle analyzes. However, one of the main factors to disregard the inclusion of these inputs is the difficulty of quantifying due to the location-dependent nature. In addition, some authors diverge between the integration or not of these inputs, since they occur independently of the agricultural system analyzed. However, it can be argued that nitrogen deposition should be included in the study, because it is a consequence of choosing the location for the agricultural system, since nitrogen deposition rates tend to vary between different agricultural management practices, soil characteristics, climatic conditions and geographic location (Brentrup, Küsters, et al., 2004).

In the literature, assessments about the use of nitrogen in agricultural systems lie in investigations regarding individual effects such as nitrate leaching or ammonia volatilization. However, agricultural production systems contribute a significant portion to a wide range of environmental impacts. These, in turn, are attributed to impact categories such as climate change (influence of nitrous oxide in the atmosphere), acidification (effect on excessive soil nitrogenation), as well as eutrophication (attributed mainly to aquatic systems). In contrast, it is important to note that the analysis of individual effects does not allow a general conclusion from an environmental point of view on the general preference of one or another production strategy.

From the same perspective, although most of the focus related to nutrient management and environmental pollution has been nitrogen during the past few years, other nutrients also need attention. Phosphates, in addition to being chemical substances with potential contaminating characteristics that cause eutrophication in freshwater bodies, these substances are recognized for presenting themselves as a limited resource. Obviously, both nitrogenous and phosphate substances must be considered in the sustainable management of agricultural systems. However, potassium is not an important source of contamination of natural waters or a limited resource. Besides, under certain conditions, a lack of sufficient potassium for crops can be a problem. More specifically, in agricultural practices that do not use fertilizers such as phosphorus or potassium, as well as not using animal fertilizers, there is an obvious risk of developing deficiencies. This is also true for essential nutrients, although many soils have adequate reserves of climatic conditions in the near future (Spångberg, Tidåker, & Jönsson, 2014).

At the same time, the same arguments are used to quantify heavy metals. However, in this case, in particular, these substances have the potential for absorption by cultures. That is, the grains of agricultural crops generally retain these substances and can subsequently contribute to different degrees of human toxicity. Thus, the assessment of impacts is not limited to the average environmental indicators but can also result in data at the endpoint and therefore offer a particular risk to human health. Thus, this specific toxicity occurs due to the agricultural system under analysis and, therefore, is relevant for inclusion. However, in many studies reported in the literature, the atmospheric deposition of heavy metals is not included due to a lack of data.

In addition, agrochemicals include pesticides and fertilizers. Conceptually, the pesticide is a general term that covers all chemicals (for example, insecticides, fungicides, herbicides, or others) used to protect the farm or product from different pests, diseases, or unwanted plant growth. Global estimates indicate that both the production of pesticides and fertilizers contributes to the potential for environmental impact at the place of production. In addition to its beneficial effects on crop growth and yield, the application of agrochemicals not only results in emissions from the field but also expands to the buildings in which animals are kept, as well as where families reside. Pesticides are considered mainly because of their toxic impacts on nontarget species during or after application and bioaccumulation on plants harvested from crop parts, potentially contributing to effects related to ecotoxicity and human toxicity. As pesticides are distributed in the agricultural field when used, the disposal stage is generally not included in LCAs. However, pesticide residues are present in the packaging materials, in the sprayer and in the water used to clean the sprayer after application. Thus, ideally, the disposal and subsequent destination of these residues should be included in LCA studies (Dijkman, Basset-Mens, Antón, & Núñez, 2018).

However, often neglected in agricultural LCAs, the mineral balance must be used to ensure that total input equals total production. For this to be possible, a steady-state must be assumed. In other words, we assume that the technical coefficients in the case study will remain the same in the future. This does not mean that all parameters will remain stable: for example, the uniform application of a certain amount of fertilizer can lead to a (constant) accumulation of heavy metals. This assumption of steady-state applies not only to wheat but also to any crop or crop pattern studied. How realistic this assumption remains is a problem for further investigation.

Faced with this problem, policy-makers, as well as the union of researchers and society, resulted in the development of a specific tool for quantifying the indicators of environmental impacts correctly for the agricultural and food sector. Thus, to ensure that the measures are reliable without inconsistency in the data collected, the SALCA project was created.

More specifically, the Agroscope life cycle assessment method, SALCA (Swiss Agricultural Life Cycle Assessment), was established to offer a method adapted to agricultural conditions to all interested parties. It comprises our developments for specific categories of emissions and impacts, as well as a variety of impact assessment models established to apply LCA to agricultural processes. Developed initially for Switzerland, the method has also been extended, so far, to neighboring countries, so that SALCA is also used in European projects.

This tool has four significant groups of characterization factors for a rigorous assessment, namely the categories of pesticides, soil quality, terrestrial biodiversity, heavy metals, nitrates, and phosphates. Below, a synthesis of each category will be described.

For the category of pesticides, the SALCA method addresses the characterization factors of the active ingredients of the pesticides were calculated according to the USES-LCA and EDIP 97 method as part of the ENDURE project (European Network for the Durable Exploitation of Water Protection Strategies), funded by the European Union under the Sixth Framework Program. The data necessary for the characterization were made available by the University of Hertfordshire and the Julius Kühn Institute.

The soil quality impact category SALCA uses nine direct indicators that express the effect of crop rotation on the soil (soil depth, the volume of macropores, aggregate stability, humus content, heavy metal content, organic pollutants, microbial activity, earthworm, and microbial biomass). Indirect indicators that can be derived from current management practices are used to determine these direct indicators. The model was designed to record long-term changes in soil quality.

The SALCA terrestrial biodiversity impact category covers 11 indicators (grassland flora, segetal flora, birds, mammals, amphibians, mollusks, bees, spiders, terrestrial beetles, butterflies, and grasshoppers). The biodiversity potential of each indicator in terms of species diversity and composition is determined in two stages: first, agricultural activities are classified based on their effects on organisms. In the second step, the defined habitat types are weighted according to their importance for the indicators. Multiplying the ratings of steps one and two yields the final score.

For SALCA heavy metals, a method was developed to record heavy metal flows, taking into account the heavy metals cadmium, copper, zinc, lead, nickel, chromium, and mercury. Emissions to agricultural soil, surface, and groundwater are determined, taking into account the input of heavy metals from seeds, fertilizers, plant protection products, and animal feed, as well as a deposition. An allocation factor is used to distinguish between diffuse inputs and those caused by agriculture. The method uses generic data for Switzerland.

Nitrate leaching by the SALCA method is defined as the surplus between supply (mineralization and fertilization) and absorption by the plant. The model takes into account both green manure and grassland. In addition, leaching is not calculated for any crop during months with low rainfall and high growth rates. The nitrate remaining at the end of this period is considered to be at risk of leaching. The method is applicable to the central Swiss plateau and is subsequently being developed for neighboring countries. Likewise, Phosphorus SALCA records phosphorus emissions taking into account losses from soil erosion, runoff, and leaching into groundwater.

In fact, both the LCA tool and SALCA are methodologies that help in identifying the environmental issues that present the most significant problems in view of the consolidation of the systems' sustainability. Only with the application of these methodologies will it be able to provide a solid basis for the development of policies that move agriculture a step closer to environmental and economic sustainability.

However, regardless of identifying and quantifying environmental impacts, the best action against these charges is to apply strategies to deal with the excessive demand for nutrients in agricultural systems. Given this scenario, three options can be used alone or in combination. The first is to reduce inputs to balance nutrient exports from the land area, for example, improved feeding strategies, including increased use of nutrients. In a second step, increase the land area for manure application. That is, reflect on the possibility of buying or renting more land or contract with neighboring farms to receive excess manure, exporting liquid manure is generally limited by the economy to an area 15 km from the farm. And, finally, export the farm's excess nutrients in the form of value-added products. This requires three conditions: the products must be odor-free, free of pathogens, and dehydrated, which essentially means that the entire volume must be processed, either by composting or by anaerobic digestion.

So, it is important to note that balanced nutrient inputs and outputs are of the utmost importance to achieve long-term sustainability in agricultural systems. The introduction of farm-wide nutritional accounting is an important tool for directing nutrient management on the farm. Besides, we need to avoid the depletion of essential nutrients, as well as the accumulation of toxic or polluting elements. In this way, when the balance is reached, we have a better chance of providing crops with sufficient essential nutrients while avoiding large losses that can cause environmental disturbances.

However, it is crucial to keep in mind that there is no single solution for obtaining sustainable agricultural systems concerning nutrients. At the farm level, nutrient management is controlled by government schemes and private initiatives. To avoid loss of nutrients on larger scales, spatial heterogeneity in the landscape must be considered; therefore, we use the most appropriate countermeasures for each field or part of an area. Therefore, a rigorous effort must be made to increase large-scale nutrient recycling to reduce local/regional excesses of nutrients from livestock manure, as well as those in the human food chain. Thus, these recycled nutrients can effectively replace part of the large number of fertilizers used in agricultural production and help to stabilize or reduce the overall levels of nutrients introduced into the environment.

Other parameters for agricultural systems

It is known that exchanges of specific gases in soils, such as carbon dioxide (CO_2), can significantly contribute to the global warming potential (GWP) of agroecosystems. Thus, due to variations in soil type, climatic conditions and land management practices, the exchange of CO_2 can differ markedly in different geographic locations. Faced with this scenario, the food industry is developing carbon footprints for its products, requiring the integration of soil CO_2 exchange with other CO_2 emissions along the food chain. This measure may have advantageous characteristics when cultivating certain crops in different geographical locations to minimize CO_2 emissions from the soil, and this may offer the potential to offset other emissions in the food chain, such as transport (Dold et al., 2017).

The agroecosystem's LCA is generally applied to assess and compare the impact of conventional and organic agriculture on environmental sustainability. The main factors of environmental impacts and climate change include CO_2 respirational emissions (Yao, Huang, & Song, 2019), assimilated and sequestered CO_2 in plants (CPE), the net production of the carbon ecosystem (NEP), and the emissions of applied chemicals and agrotechnical means throughout the harvesting system, from sowing to harvesting (from the cradle to the gate). The effects of a given crop within a cropping system are highly dependent on previous cultures (Koerber et al., 2009).

In order to account for data referring to the life cycle inventory (LCI), the parameters consist of two main mandatory elements, which are the ecological impacts, that is, gross primary production (GPP, $\mu mol \times m^2 \times s^1$), net production of ecosystems ($\mu mol \times m^2 \times s^1$), and respiratory emissions from soil and autotrophs (R_{s+a}, $\mu mol \times m^2 \times s^1$). And, subsequently, the second mandatory element is associated with energy-based impacts, that is, labor input (E_{da}) together with applied agrochemicals (E_{dt}).

Among the mandatory elements, net ecosystem production (NEP) is used to represent carbon exchange and is defined according to Eq. 7.2:

$$NEP = GPP - R_{a+s} \tag{7.2}$$

where NEP is the difference between gross primary production and emission of soil breath and autotrophic ($\mu mol \times m^2 \times s^1$), GPP is the gross primary production ($\mu mol \times m^2 \times s^1$), and R_{a+s} is the respiratory emissions from soil and autotrophs ($\mu mol \times m^2 \times s^1$).

The flow of CO_2 has been converted into the carbon equivalent (C), and net ecosystem production (NEP), ecosystem respiration (R_{ECO}), and gross primary production (GPP) can be computed as:

$$NEP_d = F_{Cday} + F_{Cnight} \tag{7.3}$$

$$R_{ECOd} = R_{day} + F_{Cnight} \tag{7.4}$$

$$GPP_d = NEP_d - R_{ECOd} \tag{7.5}$$

where NEP_d is the daily net ecosystem production ($gC \times m^{-2} \times d^{-1}$), R_{ECOd} is the daily ecosystem respiration ($gC \times m^{-2} \times d^{-1}$), GPP_d is the daily gross primary production ($gC \times m^{-2} \times d^{-1}$), F_{Cday} is the daily sum of daytime carbon flux ($gC \times m^{-2} \times d^{-1}$), F_{Cnight} is the daily sum of nighttime carbon flux (i.e., nighttime respiration) ($gC \times m^{-2} \times d^{-1}$), R_{day} is the daytime respiration, either from F_{Cday} (during the off season) or calculated with Eq. (7.3) ($gC \times m^{-2} \times d^{-1}$).

It is important to note that the arithmetic convention of the signals is negative for flows from the surface to the atmosphere and that the NEP can be less than zero. The annual sums of NEP and GPP can be computed according to Eqs. (7.6)–(7.8):

$$NEP_{year} = \frac{\sum NEP_d}{d \times n_{year}} \tag{7.6}$$

$$R_{ECO_{year}} = \frac{\sum R_{ECOd}}{d \times n_{year}} \tag{7.7}$$

$$GPP_{year} = \frac{\sum GPP_d}{d \times n_{year}} \tag{7.8}$$

where NEP_{year} is the annual difference between gross primary production and emission of soil breath and autotrophic ($\mu mol \times m^2 \times year^{-1}$), $R_{ECOyear}$ is the annual ecosystem respiration ($gC \times m^{-2} \times year^{-1}$), GPP_{year} is the annual gross primary production ($\mu mol \times m^2 \times year^{-1}$), NEP_d is the daily net ecosystem production ($gC \times m^{-2} \times d^{-1}$), R_{ECOd} is the daily ecosystem respiration ($gC \times m^{-2} \times d^{-1}$), GPP_d is the daily gross primary production ($gC \times m^{-2}$ - $\times d^{-1}$), d is the number of days measured (d) or filled with intervals, and n_{year} is the number of days per year (d).

On the other hand, the life cycle inventory based on exergy results in the estimation of CO_2 emissions equal to the labor input (E_{da}), together with the applied agrochemicals (E_{dt}), calculated according to Eq. 7.9:

$$E = E_{da} + E_{dt} \tag{7.9}$$

where E is the exergy results in the estimation of CO_2 emissions (kg CO_2 eq), E_{da} is the labor input (kg CO_2 eq), and E_{dt} is the emissions of applied agrochemicals (kg CO_2 eq).

However, the carbon content in the crop yield can be estimated by Eq. 7.10:

$$C_y = DM \times CF \times 0.01 \tag{7.10}$$

where C_y is the carbon yield (ton/ha^{-1}), DM is the dry biomass per unit area (ton/ha^{-1}), CF is the carbon fraction (ton/t^{-1} dry matter), and 0.01 is the correction factor. In the analysis, we can assume the carbon fraction of 0.43 ($Cton/ton^{-1}$ DM).

The total carbon balance (C_b) used as a complete parameter for assessing LCA is defined as the difference between the emitted CO_2 obtained from the LCI (t/ha^{-1}) and the CO_2 accumulated in the NEP, and removed with the harvest yield (C_y, ton/ha^{-1}) in organic and conventional agroecosystems, on accordance with Eq. 7.11:

$$C_b = E - NEP - C_y \tag{7.11}$$

where C_b is the total carbon balance (t/ha^{-1}), E is the emitted CO_2 from LCI (ton/ha^{-1}), C_y is the removed harvest yield (ton/ha^{-1}).

However, it is important to highlight that negative Cb values indicate net carbon sequestration from the atmosphere, while positive values represent net carbon emissions from the agroecosystem to the atmosphere. Besides, this carbon footprint can be used in studies as a standard indicator of environmental impact within the LCA methodology.

References and recommended reading

Audsley, E., Alber, S., Clift, R., Cowell, S., Crettaz, P., Gaillard, G., … Pearce, D. (1997). *Harmonisation of environmental life cycle assessment for agriculture*. Final Report, Concerted Action AIR3-CT94-2028 (p. 139). European Commission, DG VI Agriculture.

Bergström, L., Bowman, B. T., & Sims, J. T. (2005). Definition of sustainable and unsustainable issues in nutrient management of modern agriculture. *Soil Use and Management*, *21*(1), 76–81.

Brentrup, F., Küsters, J., Kuhlmann, H., & Lammel, J. (2004). Environmental impact assessment of agricultural production systems using the life cycle assessment methodology: I. Theoretical concept of a LCA method tailored to crop production. *European Journal of Agronomy*, *20*(3), 247–264.

Brentrup, F., Küsters, J., Lammel, J., Barraclough, P., & Kuhlmann, H. (2004). Environmental impact assessment of agricultural production systems using the life cycle assessment (LCA) methodology II. The application to N fertilizer use in winter wheat production systems. *European Journal of Agronomy*, *20*(3), 265–279.

Dijkman, T. J., Basset-Mens, C., Antón, A., & Núñez, M. (2018). LCA of food and agriculture. In *Life cycle assessment* (pp. 723–754). Cham: Springer.

Dold, C., Büyükcangaz, H., Rondinelli, W., Prueger, J. H., Sauer, T. J., & Hatfield, J. L. (2017). Long-term carbon uptake of agro-ecosystems in the Midwest. *Agricultural and Forest Meteorology*, *232*, 128–140.

Englund, O., Börjesson, P., Berndes, G., Scarlat, N., Dallemand, J. F., Grizzetti, B., & Fahl, F. (2020). Beneficial land use change: Strategic expansion of new biomass plantations can reduce environmental impacts from EU agriculture. *Global Environmental Change*, *60*, 101990. https://doi.org/10.1016/j.gloenvcha.2019.101990.

Koerber, G. R., Edwards-Jones, G., Hill, P. W., i Canals, L. M., Nyeko, P., York, E. H., & Jones, D. L. (2009). Geographical variation in carbon dioxide fluxes from soils in agro-ecosystems and its implications for life-cycle assessment. *Journal of Applied Ecology*, *46*(2), 306–314.

Mackenzie, S. G., Leinonen, I., & Kyriazakis, I. (2017). The need for co-product allocation in the life cycle assessment of agricultural systems—Is "biophysical" allocation progress? *The International Journal of Life Cycle Assessment*, *22*(2), 128–137.

Mourad, A. L., Coltro, L., Oliveira, P. A., Kletecke, R. M., & Baddini, J. P. O. (2007). A simple methodology for elaborating the life cycle inventory of agricultural products. *The International Journal of Life Cycle Assessment*, *12*(6), 408.

Pawelzik, P., Carus, M., Hotchkiss, J., Narayan, R., Selke, S., Wellisch, M., … Patel, M. K. (2013). Critical aspects in the life cycle assessment (LCA) of bio-based materials—Reviewing methodologies and deriving recommendations. *Resources, Conservation and Recycling*, *73*, 211–228.

Scherer, L., van Baren, S. A., & van Bodegom, P. M. (2020). Characterizing land use impacts on functional plant diversity for life cycle assessments. *Environmental Science & Technology*, *54*(11), 6486–6495.

Spångberg, J., Tidåker, P., & Jönsson, H. (2014). Environmental impact of recycling nutrients in human excreta to agriculture compared with enhanced wastewater treatment. *Science of the Total Environment*, *493*, 209–219.

Tsangas, M., Gavriel, I., Doula, M., Xeni, F., & Zorpas, A. A. (2020). Life cycle analysis in the framework of agricultural strategic development planning in the Balkan region. *Sustainability*, *12*(5), 1813.

Venter, Z. S., Scott, S. L., Desmet, P. G., & Hoffman, M. T. (2020). Application of Landsat-derived vegetation trends over South Africa: Potential for monitoring land degradation and restoration. *Ecological Indicators*, *113*, 106206. https://doi.org/10.1016/j.ecolind.2020.106206.

Yao, X., Huang, R., & Song, M. (2019). How to reduce carbon emissions of small and medium enterprises (SMEs) by knowledge sharing in China. *Production Planning & Control*, *30*(10 – 12), 881–892.

Reidsma, P., Tekelenburg, T., Van den Berg, M., & Alkemade, R. (2006). Impacts of land-use change on biodiversity: An assessment of agricultural biodiversity in the European Union. *Agriculture, Ecosystems & Environment*, *114*(1), 86–102.

Chapter 8

Economic aspects of sustainability

The economic science presents many obstacles to be elucidated. The economic aspects are based on very volatile theories. This is because, for each theory on economics, another one appears almost instantly, contradicting it. Thus, in its most simplified definition, the sustainable economy is based on how we should allocate financial resources to meet socioeconomic needs, in addition to social well-being. However, economists hardly agree with each other as to which way is the most appropriate possible to dispose of the resources.

On the other hand, the impasse posed by economists is far from being resolved. Although countless scientists design scientific designs to integrate economic fundamentals with sustainable aspects, in fact, for the most part, the tremendous economic echelon continues to contribute so that this understanding remains a complicated outcome.

For many years, the main governmental challenge was to combat socioeconomic differences and, at the same time, guarantee economic growth. On the other hand, perhaps with even greater complexity, today, the obstacle is another. At present, economic progress has required significant efforts for the transition to be more excellent. In fact, to further complicate matters, we are now trying to achieve economic growth while protecting the environment at the same time. Consequently, the fear was installed, and a major question needs to be answered: How to reconcile economic aspects in the face of sustainability? Will the business circles of the value chains of large industrial and agricultural segments manage to set aside their convergences and install management measures in which economic success goes hand in hand with the environmental viability of processes and products?

Because of these questions, first, it is necessary to understand some basic definitions in order to proceed in an attempt to understand the complexity of this theme. In fact, the literature reports two ways of referring to economic sustainability. In theory, the first has as an elementary concept that organizations remain in the enterprise and approach the issue from the inside out. That is, there is a whole process of looking more closely at the flows of operations and investments from within the industry. However, the second sustainable economic

Sustainability Metrics and Indicators of Environmental Impact. https://doi.org/10.1016/B978-0-12-823411-2.00002-5

model has the main focus on the economic impacts that the organization causes on society, that is, the external view of the interested parties. Generally, this vision is guided by how society recognizes its environmental efforts, through sustainable labels, or even in the profits that the appeal of its products is capable of providing.

Regardless of which direction to use, undeniably economic sustainability in its most upright format requires us to reconsider both the internal and external implications of sustainable management. Thus, this means that some essential criteria must be considered to ensure the management of economic sustainability. For this, it is necessary to adopt measures such as assessing the performance of the financial performance of the segment, more precisely how the administrator is able to conduct intangible assets. In addition, the profile of conduct and its influence on the economy, in general, and how it will influence management and environmental and social impacts.

In addition, more considerable attention must be paid to the inner vision. This is because this measure aims to assess the main bottlenecks related to corporate turnover and the reputation of the brand to be cared for. Thus, greater depth must necessarily be explored when considering economic sustainability as the center of the process. Therefore, although this orientation does not report the global context, they believe it roughly a critical part. Thus, the gap to be filled encompasses more precise issues regarding sustainability management tools. And only after this discovery, will it be possible to deal freely and without prejudice about the economic impacts that a follow-up has on society. This is because, just as it is imperative for businesses to generate profits, companies must remain in business to create sustainability as a whole.

On the other hand, if many fail to detach themselves from the negative aspects that profit generation can provide, exclusive advocates of socioenvironmental criteria seem to see no reason in this matter. Therefore, another question needs to be answered. Why is economic sustainability important?

To meet the need for an answer, the literature has brought several studies highlighting the importance of a sustainable economy. According to some studies projected by the New Economics Foundation, a prominent project received notorious attention to questions of how the economic aspects of sustainability should be explored to achieve a more comprehensive understanding of its implications for sustainability as a whole (Doane & MacGillivray, 2001). Its main objective was to identify the ways to capture the principles in a great management structure present within the administrative sectors (Blühdorn, 2017).

According to this project, it was possible to identify viable hybrid methodology, quantifying process inputs and outputs, and feasible strategic measures in which improvements in economic performance could be applied, considering the short- and long-term risks associated with environmental indicators.

However, there is a consensus among the traditional socioenvironmental ones whose estimate is only the segments that manage the environment responsibly. Leaving, as a second option, those that provide transparent socioeconomic benefits to the community in which they are operating. Besides, in a sustainable economy, only the best must and will survive.

Given this context, industrial and agricultural segments that have strategies that employ social and environmental sustainability as the focus of their actions, while their profit remains expressive, will have their success recognized. However, from another point of view, we will still face probable business segments that tend to follow the opposite movement and taking advantage of the concepts that emerged by sustainable actions. For this reason, in many cases, green appeals can be seen as dubious opportunities (Jänicke, 2012).

On the other hand, it is known that for a long time the fundamentals and justifications regarding the criteria taken into account to stand out in social and environmental performances were developed. However, they are not yet convincing. In short, for many managers, just knowing the criteria listed, does not make the processes economically sustainable.

To that end, some approaches taken by the Institution of Chemical Engineers and the United Nations used four main criteria to define important parameters for the economic dimension, namely (1) financial health, (2) economic performance, (3) potential financial benefits, and (4) trading opportunities. Faced with this, the main objective of the research evidenced financial stability and short- and long-term survival skills in business.

In a brief conceptual explanation, financial health can be understood as the criterion that comprises the aspects that analyze the solidity of the processes. In other words, the internal financial stability of the industrial and agricultural segments must contain classic subcriteria such as profitability, liquidity, and solvency.

In contrast, the economic performance aims to assess the company's value as perceived by external shareholders. That is, senior management and the government have the most relevant influences since the subcriteria associated with this criterion include share profitability, gross domestic product contributions, in addition to performance and market share.

On the other hand, potential financial benefits assess financial benefits that do not involve profits more carefully. In terms of exemplification, national and international subsidies based on environmental, social, and technological improvements can be framed like this model. To that end, some initiatives should be proposed to project projects potentially eligible for funding. A potential example that has been suggested a lot is the clean development mechanism (CDM) agreed under strict regulations of the nations like the Kyoto Protocol (see Chapter 6).

Finally, the design of negotiation opportunities aims to examine the vulnerability of the company's commercial network, as well as the risks to which it is exposed by the system in which it is inserted, considering the number of national or international companies in the commercial network.

As can be seen, the criteria taken into account are perceived from the perspective of conventional economists. Therefore, only one approach, in fact, considers sustainability as a relevant criterion. In addition, economists notoriously lack the theoretical or empirical tools to predict the impact of economic activities on the environment. Thus, some perspectives suggest that sustainable development is seen as an element of sustainable economic development. Therefore, this also creates confusion about sustainability and economic viability, that is, sustained growth and its self-reliance.

As noted, these conceptions have minimal design. They often neglect the impacts of economic activities on the environment, as well as the possible damage associated with these activities, which can influence the community in the short and long terms. This, in turn, is the most significant dilemma found in this topic, much in consideration for the fact that environmental damage should not be expressed in monetary terms. Likewise, under an ethical stance, the value of societies and their individuals cannot be priced either. Equally, this bottleneck to be overcome tends to use compensation strategies, through governmental models, to rate the possible impacts caused to society and ecosystems.

Therefore, at first, what can be concluded is that, in the short term, modern ideologies depend on economic profits. In addition, scientific traditions based on reductionist cause and effect relationships are failing to explain and address the complex dynamic interrelationships between economic, environmental, and social aspects and the perspective of time. Thus, in the following, we will elucidate a little more about the main mechanisms and theories described in the arguments that plague sustainability in the face of economic aspects.

An approach to the economic aspects of corporate institutional sustainability

Meeting the economic issues of an institution is not easy. However, achieving the economic sustainability approach is an even more challenging task. Therefore, numerous research structures incorporated in the front line of the most diverse industrial segments are constantly concerned with how to assertively manage the impacts associated with all stages of production line processing and/or activities performed whose economic circumstances of its stakeholders and their economic systems at all levels of global society may be affected (Baumgartner, 2009).

Under the same view, the capital generated in the establishment can be interpreted in different ways. And, therefore, economists can focus on several types, being artificial, natural, human, social whose must always be sustainable. For this purpose, administrators generally propose theories that are more predisposed to corporate marketing. According to the description of the United Nations, the approach to institutional economic sustainability tends to be at a more strategic level.

In view of this scenario, the first criterion aims to mention and incorporate the principles of sustainability in business strategies. For this, the development of marketing about visibility has as main objective the elaboration of a strategic plan. In this sense, the first step for a follow-up to be well directed, it is essential to have a well-defined mission, vision, and values. Briefly, a company's mission is a concise statement of the organization's fundamental purpose, the purpose of its existence, the reason why it was created. The mission is like the DNA, defining its identity and does not usually change over time.

In contrast, a company's vision represents a future state for the business, where it wants to go, what it wants to achieve. Unlike the mission, the company's vision is created for a predetermined period so that the vision can change over time, according to the moment of the organization. Finally, the values can be understood through the principles or beliefs that serve as a guide for the behaviors, attitudes, and decisions of all people involved in the pursuit of their objectives so that the segment can carry out its mission and seek to achieve its vision. Therefore, these indicators must be aligned with those of national and international governments.

Subsequently, the second criterion can be understood by openly recognizing support for global agreements. This is because global agreements have ambitious objectives to preserve biodiversity, whose goals range from preventing species extinction to decreasing deforestation rates. Among the main global deals, we can highlight the Paris Agreement, the United Nations Conference on Climate Change (COP), and the Nagoya Protocol. However, it is essential to rethink the link and the form of these agreements. Generally, strict compliance is not very useful. These are nonbinding commitments, which countries do not take seriously. Therefore, it is expected that global agreements, in fact, will have importance when decision makers, public and private, understand that continuing to act as they do today is not an option for nature.

In addition, the third criterion is intended to include external sustainable development objectives in internal research and development. This criterion is related to the guidelines and relationships between political-governmental managers (external forces) and decision makers (internal strength) to achieve sustainable, inclusive, and sustained economic growth. Besides, it requires ensuring the significant mobilization of resources from a variety of sources, including through strengthening development cooperation to provide adequate and predictable means through the environmental connection.

Finally, the fourth criterion aims to allocate funds to address sustainability issues beyond the industrial segment's immediate control. For this, the corporation's environmental management must act consciously, always combining respect for the environment and the society in which it is inserted. Thus, it is essential to understand how sustainability should be adopted and implemented. Therefore, it must allow the business to sustain itself, survive, even choosing to act sustainably.

Strategies for the manifestation of the economic aspects of institutional sustainability

As seen previously, economic sustainability encompasses several general aspects of an organization that need to be respected, in addition to environmental and social aspects, to maintain the competitiveness of the industrial segment. However, to achieve financial stability and liquidity, profitability, and financial benefits resulting from sustainability activities, it is necessary to implement primary elements that make up the economic dimension (Table 8.1). Thus, this incorporation is achieved through management strategies, which provide the basis for the definition of a corporate sustainability strategy (Schneider, 2015).

The manifestation of institutional sustainability at a strategic level within a business or industrial and agricultural segments can, therefore, be seen as a prerequisite for sustainable operations, projects, or even corporate sustainability. The first level of the proposed sustainability assessment framework is consequently called the corporate responsibility strategy (Baumgartner, 2014). This implies that it is a prerequisite for all sustainability, and it is a strategy that accepts the company's responsibility and its vital role in all societies in which it operates, and also in the global environment (Fig. 8.1).

There are different types of sustainability strategies. Generally, categorical models are the most applied in industrial segments. Conceptually, these models are identified as those that allow the identification and structuring of characteristics of strategic environmental management at a given moment or continuum/progression models that integrate the evolution of strategies over time. However, for a comprehensive corporate sustainability strategy, it is necessary to consider all dimensions of sustainability, its impacts, and interrelations (Baumgartner & Rauter, 2017). In addition, it is possible to distinguish different sustainability strategies based on a variety of reactive approaches to an offensive and proactive strategy, in this way they can be interpreted as simultaneous progression and as categorical models.

Given this context, Table 8.2 summarizes the main strategies applied to the economic aspects of sustainability.

Simply put, these types of strategies describe generic possibilities for dealing with the sustainability challenge, that is, with different environmental and social aspects of commercial activities, according to the principles of sustainability. However, the choice of the applied strategy will have different impacts on the global perspective. For example, strategies with an external focus (extroverted and visionary sustainability strategy) are likely to have a particularly strong social impact. When achieving improvements in terms of the sustainability issues described earlier, internal and external stakeholders (employees, customers, consumers, suppliers, neighbors, and society itself) are likely to benefit. On the other hand, an extroverted strategy can be transformative. This can be explained once a transformative strategy interacts with the market and actively tries to change market conditions. Thus, this strategy aims to create new market

TABLE 8.1 Economic aspects of corporate sustainability.

Parameter	Conceptualization
Innovation and technology	Efforts made in R&D related to sustainability to reduce the environmental impacts of new products and business activities. Use of best available techniques (BAT) and integrated environmental technologies, focusing on cleaner production and zero-emission technologies.
Collaboration	Excellent cooperation and active collaboration with various partners (for example, suppliers, R&D institutions, universities). Working on shared programs and networks to develop innovative products and technologies. Exchange of information and knowledge.
Knowledge management	Activities and approaches that maintain knowledge related to sustainability in the organization. Methods for planning, developing, organizing, preserving, transferring, applying, and measuring specific expertise and improving the organizational knowledge base.
Processes	Transparent processes and functions are defined so that business activities are conducted efficiently, and every employee knows what the organization expects of them (also concerning sustainability). Adaptations of process management to achieve sustainability require the systematic implementation of corporate sustainability—integration of sustainability in daily business life.
Purchase	Consideration of issues related to sustainability in purchases. Awareness and consideration of issues related to sustainability in the organization, as well as in the entire supply chain. Relationship with suppliers, also focusing on sustainability.
Sustainability reports	Inclusion of issues related to sustainability in company reports, either in individual sustainability reports or integrated into corporate reports.

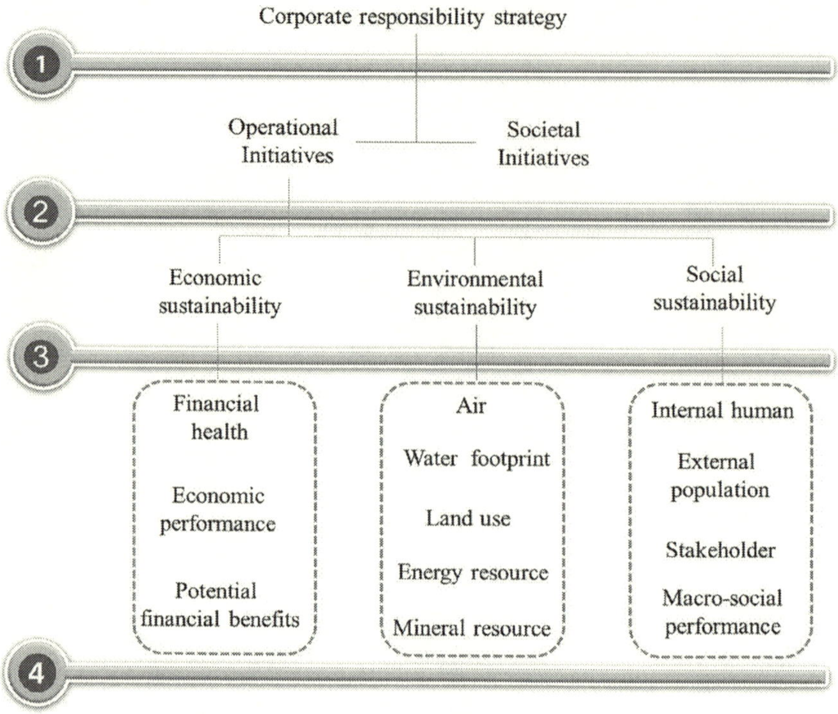

FIG. 8.1 Leveling of the proposed structure for economic sustainability.
This figure shows an example of a strategy applied to the corporate structure integrating the pillars of sustainability. *(Based on Tharp & Chadhury, 2008.)*

TABLE 8.2 Main strategies and their respective definitions of sustainable economic aspects.

Strategy	Focus/definition
Mitigation of introverted risks	Focus on legal norms and other external norms related to environmental and social aspects, in order to avoid risks to the company.
Extroverted—legitimizing	Focus on external relationships, licenses to operate.
Conservative—efficiency strategy	Focus on eco-efficiency and cleaner production.
Visionary	Focus on sustainability issues in all business activities; competitive advantages are derived from differentiation and innovation, offering unique benefits to customers and stakeholders.

opportunities in the light of sustainable development, including elements of the conservative and visionary approach (Baumgartner & Ebner, 2010).

At the same time, visionary strategies occur in two different ways: conventionally and systemically. Conventional visionary strategies are based on market opportunities in an opportunistic way. As long as sustainability issues lead to market advantages, they are part of the strategic management of conventional visionary-oriented companies. The focus is from the outside in; the inputs for formulating the strategy are derived from the market perspective. Systemic visionary strategies combine this vision with an inside-out perspective, that is, the marker-based view is complemented with a resource-based view, and sustainable development is deeply rooted in the company's regulatory level.

Notoriously, the vision of strategies to be incorporated can be understood as an empty approach and without much veracity. This view is the result of the lack of guidance spent by these models and therefore is one of the main reasons for the lack of progress in this field. However, bone can be compensated to a certain extent, clarifying the respective opportunities, benefits, risks, and trade-offs associated with the implementation of corporate sustainability.

In addition, although strategic management in the discussion on corporate sustainability is considered unfinished, there is a consensus among decision makers whose model should allow companies to create commercial and social value through their activities related to sustainability. In this way, they suggest not only theoretical propositions but also provide concrete examples of how these propositions could be applied in practice.

An approach to the economic aspects of sustainability from the perspective of the shared economy

It is widely disseminated that sustainable consumption reduces the impact on the environment so that human needs can be met not only in the present but also for future generations. To achieve this maxim, the advances and discoveries current in technological innovations give rise to the formation of increasingly consolidated ties with consumers in the search to understand and elucidate the fundamental parameters that guarantee the economic sustainability of industrial and agricultural segments (Dabbous & Tarhini, 2019).

Faced with this, the perspective of the shared economy through technology, mainly through online providers, aims to provide an economic system in which goods or services are shared with the intention of making efficient use of social resources that encourage the change to a more sustainable mode of consumption. Thus, this commitment to society and industrial follow-up have numerous advantages, among them, this movement helps to share unused resources with others to reduce waste and, eventually, leads to a higher level of common interests in society.

However, in practice, this concept external to the main process flows present within the industrial segment, where remedies are punctually displayed, can be applied. On the other hand, the treatments used in the proceedings must not present intolerable residual risks. For this, the maximum premise attributed is to save money without increasing the use of natural resources. Through this theory, the sustainability assessment is assigned to assess, in fact, the reduction of environmental impacts, generally conducted by carbon footprint indicators together with cost-benefit analyzes (CBA).

Cost-benefit analyzes have different cost and benefit structures and items to be accounted for. However, generally, attributes such as the costs of carrying out the correction, as well as negative externalities due to the corrective action, that is, costs not incurred by the party responsible for the correction, are determined. The main economic indicators considered to assess economic sustainability are the cost of capital, total annualized costs, and capped costs (Santoyo-Castelazo & Azapagic, 2014). However, to carry out a more in-depth economic analysis, details such as calculations related to companies' investment modalities, such as CAPEX (expenses with the acquisition of goods) and OPEX (operating costs), are necessary. Nonetheless, the specific details of CAPEX and OPEX are not the chapter objective; therefore, more details about the content can be found in classical literature, such as (Peters, Timmerhaus, and West (2003)). Thus, since energy demand is generally the main problem attributed to the production systems, later, we will direct, in terms of example, quantitative energy parameters associated with the economic aspects of sustainability.

Briefly, capital costs include the costs of building and installing industrial segments. They are calculated as overnight costs, that is, interest-free costs on loans, and can be calculated according to Eq. (8.1):

$$TC_C = \sum C_C \times E \tag{8.1}$$

where TC_C is the total cost of capital (USD), C_C is the overnight capital costs of the energy generation option (USD/kW), and E is the installed capacity of the industrial segment (kW).

In addition, the second indicator can be attributed to the total annualized cost of an energy system. This, in turn, is defined by Eq. (8.2):

$$TAG = \sum AC_C + \sum F_C + \sum V_C + \sum f_C \tag{8.2}$$

where TAG is annualized total cost of energy generation (USD/year), AC_C is annualized cost of capital (USD/year), F_C is the annual fixed costs (USD/year), V_C is annual variable costs (all variable costs, excluding fuel costs) (USD/year), and C annual fuel costs (USD/year).

In contrast, annualized capital costs (AC_C) are calculated taking into account the total cost of capital (TC_C) and an annuity factor (f), as shown in Eqs. (8.3) and (8.4):

$$AC_C = TC_C \times f \tag{8.3}$$

$$f = \frac{z(1+z)^t}{(1+z)^t - 1} \tag{8.4}$$

where TC_C is the total cost of capital (USD), z is the discount rate, and t is the useful life of the plant (years).

In addition, the annual fixed costs (AFC) comprise the costs of operating an active segment for more than a year and include costs of operating personnel, insurance, taxes, repairs, or spare parts costs. Annual variable cost (AVC) annual variable costs include expenses related, for example, to contracted staff, consumed materials, and operating waste disposal costs per year, excluding fuel costs. Annual fixed costs fuel represents the cost of fuels consumed in electricity production per year.

On the other hand, if the economic side has reached its positive net value, parallel parameters such as public acceptance, health, and safety, as well as intergenerational issues, will, in fact, be the real indicators of the consolidation of the industrial and agricultural segments. In this context, one of the main problems associated with public acceptability is the determination of the locations where the industry is installed. This is because the planning of the plant must take into account the aspects of the change in land use, the impact that the installations will affect between the noises and the waste generated. Similarly, risk perceptions and safety must be assessed. Thus, this scenario integrates the social pillar, a key factor to achieve the sustainability of industrial and agricultural processes, although often neglected or attributed as a subsequent indicator. Besides, intergenerational issues considered include mitigating climate change, depleting fossil fuel reserves, and managing long-term hazardous waste related to nuclear waste and carbon dioxide storage. As seen in other approaches to sustainable economic aspects, intragovernmental measures will always have an outstanding appeal in the decision making of the managers of the industrial sectors.

Subsequently, it is essential to emphasize that portraying the sustainability pattern of industrial and agricultural segments is a somewhat challenging process. This is because, among the various theories, it is worth noting to understand that in most cases environmental strategies pay financially. Following this logic, green managements propose appropriate strategies for sustainable follow-up, exploring situations where everyone wins, which reconciles environmental protection and financial success.

As a consequence, a deep-seated trend is deliberate in relation to economic performance in prioritizing financial progress. Even if there are other alternatives for improving economic performance, concurrent with environmental performance, often, the strategy of more significant financial gain will be chosen, that is, regardless of the total increase, and therefore, environmental performance is placed in the background.

Likewise, there is an established intrinsic preference for compensation cases. For example, the alternatives that come with a financial loss will always be inferior to the other options that bring financial gain, even if they result in substantial

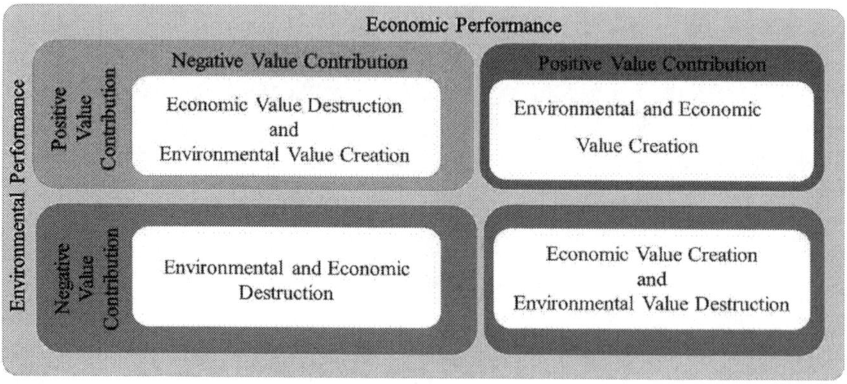

FIG. 8.2 Compensation strategies between economic and environmental creation. This figure represents a SWOT analysis model applied to economic and environmental compensation strategies. *(Based on Figge & Hahn, 2002.)*

environmental losses. Therefore, the most agreed upon measure among decision makers is to prioritize cases where everyone wins uncritically. The instances in which everyone wins are always preferred to exchange cases, even if they have a net gain more considerable than the situation in which everyone wins.

However, these views are often taken into account. Yet, they are notoriously unable to resolve the impasses of sustainable segments. To address this dilemma, this situation can only be overcome if environmental performance is analyzed separately as a complement to financial performance and not as a subordinate. With an environmental assessment based on opportunity cost similar to financial performance assessments, the economic and environmental aspects can be expressed in monetary terms. In doing so, companies can identify four different types of strategies, as shown in Fig. 8.2.

According to Fig. 8.2, compensation strategies can be divided into four major subsections, which are the strategies that bear the responsibility for integrating the use of economic capital associated with environmental resources in the generation of value. That is, they tend to generate economic and environmental value separately. On the other hand, we can also identify the strategies that lead to the use of economic capital to create value, but to the use of environmental resources that destroy value, that is, there is the creation of economic value at the expense of creating environmental value. In contrast, it is possible to observe the strategies that lead to the loss of economic value, but with the use of environmental resources that create value, that is, there is a loss of economic value, but a positive environmental value. Finally, there are strategies that do not use economic capital or environmental resources in order to create value; that is, there is a loss of economic value and negative environmental value.

From the point of view elucidated earlier, it is necessary to assign, in fact, strategies to go beyond sustainability. To achieve this purpose, inevitable efforts are required to develop a systemic vision of a marked cultural shift in paradigms among the industrial, scientific, and social community. Only then will it be possible to establish adequate conditions to support the cogeneration of an economically and environmentally sustainable system in the near future.

An approach to the economic aspects of sustainability through sustainable added value

Different from what was seen previously, the approach to the economic aspects of sustainability through sustainable added value aims to propose a monetary measure referring to corporate contributions to sustainability, being defined as the sustainable added value.

In its simplest concept, the added value is guided by the perception of considering attributes such as the efficiency of processes in terms of the absolute level, that is, the effectiveness of the use of resources. Therefore, sustainable added value can be understood as the extra value elaborated when the general level of environmental and social impacts is maintained in its constant character. In this way, it can also be associated as an alternative of potential compensation for external influences to avoid externalities, while traditional approaches compensate victims for accepting externalities.

To go deeper into this concept, first, it is necessary to understand the basic principles of measuring corporate contributions to sustainability. In this sense, we will briefly explain the two main predominant approaches: absolute measures and relative measures.

Absolute measures

Absolute measures can be understood as a way of assessing corporate contributions to sustainability. For this, it is necessary to subtract the costs from the benefits generated by the industrial and agricultural segments. This measure is intended to consider internal and external costs, that is, they may be due to the accounting of total costs. In terms of exemplification, these process systems may contribute to sustainability if the benefits exceed the sum of internal and external costs. This sum-product can be titled as the net added value or also green added value. Under these aspects, it is essential to emphasize that the economic performance of a process in terms of added value is shaped by its external environmental cost resulting from the economic activity generated by the industry. Therefore, both benefits and costs can be inferred under the same unit measure. So, it is for this reason that environmental and sometimes social damage are monetized (Figge & Hahn, 2012).

In theory, these clarifications provide a very influential basis for industrial segments. However, as in all practical problems, numerous limitations of issues are observed. In short, the most significant gap to be overcome is the requirement that environmental and social damage be monetized. This is because, as seen in the previous chapters, pricing is still misunderstood, and often its discussion generates setbacks. Therefore, the imposition of indicating environmental and social damage in monetary factors strictly hinders the practicality of absolute measures.

On the other hand, another inconvenience of these criteria consists of the comparison between the values generated by an industrial segment with the environmental and social damages caused. These approaches admit assessing whether a particular segment was able to reach the limit level of sustainability, that is, whether it is sustainable or unsustainable. However, these measures do not provide any information about the maximum contributions to be achieved. Thus, this managerial behavior influences the integration of the three dimensions of sustainability through substitution and remuneration. Therefore, absolute measures are based on the conjecture of total substitutability, being strongly influenced by weak sustainability.

Relative measures

Relative measures can be understood by the ability to express the contributions of industrial and agricultural segments to sustainability through the benefits generated by the unit of environmental/social impact. Among the most popular models of a relative measure is the determination of eco-efficiency.

Currently, eco-efficiency can be segmented into two different applications. Among them is the maximum eco-efficiency, that is, it is associated with the reduction or even the minimization of environmental impacts. On the other hand, the second perception has the term eco-efficiency to characterize the proportion of the value created by environmental impact added (EIA). It is worth mentioning that from the other understanding, when we use the term eco-efficiency (and in parallel with social efficiency), we are referring to a reason, namely, value-added environmental impact. However, in this case, the added value of an industrial segment can be fixed as the residual value that remains even after the turnover has been minimized by the demand for the cost of goods and services spent by the production process. Thus, the value of the added environmental impact represents the aggregate of all energy and material flows under consideration induced by economic activity weighted by its relative harmfulness to the environment. This aggregation is done by classifying each emission according to its contributions to different environmental problems, which are weighted by their relative harmfulness.

Under the same purpose, eco-efficiency is attributed to the duty to describe the degree to which a segment uses environmental resources in relation to its economic activity. Generally, decision makers in favor of this premise ensure that potential improvements attributed are capable of increasing the contributions of industries to sustainability. For a better explanation of the relationship between sustainability and eco-efficiency, it is plausible to blame differentiation between weak and robust improvements. For example, the strong improvements found in eco-efficiency include optimized economic and environmental performances, while weak improvements in the connection require only one dimension to be improved.

However, the bottlenecks found for the relative measures to be, in fact, consolidated need to be overcome. First, it is necessary to understand that the corresponding measures do not offer any information about the effectiveness of the processes. In short, efficiency is able to provide only information about the relationship between the consequences, that is, the possible environmental impacts and the desired final result, that is, satisfactory economic performance. From this perspective, both the economic and environmental performances of segments in absolute terms cannot be derived from eco-efficiency indexes. Thus, it is worth noting that if a given segment versus the value in monetary terms per ton of carbon dioxide, it is not possible to determine how much in monetary value that the industry created or emitted in loads of carbon dioxide in absolute terms. In this case, the absolute degree of environmental, social, and economic performances must be represented by units of effectiveness. However, as the contestation of different forms of capital in the face of substitutability is frequently observed, it is necessary to elucidate complementary references about ecological and social effectiveness. Thus, environmental effectiveness can be understood as the absolute degree of use of environmental and social resources required by industrial and agricultural segments.

Concomitantly, the progressive developments identified in environmental performance, as a result of optimized eco-efficiency, can potentially be offset by two effects. This because an improved eco-efficiency does not ensure an improved eco-efficiency. If, on the one hand, better eco-efficiency can result in growth, therefore, this process can cause a rebound effect in the use of environmental resources. In contrast, the saved environmental resources, which are the result of optimized eco-efficiency, can be used by other less eco-efficient sectors.

Besides, it is crucial to be aware that eco-efficiency does not cover social aspects. In addition, even if social issues are considered analogously as social efficiency, relative measures only allow the integration of environmental and social impacts that are expressed in the same unit. Currently, there are several accounting models used to determine relative weights for some environmental impacts, usually through analysis of the life cycle and its variations. However, to arrive at a sustainability indicator (relative or absolute), all environmental and social subindicators must be weighted and aggregated. It is very doubtful that all relevant social and environmental impacts can be integrated into a common unit of added environmental and social impacts, respectively, and thus be aggregated into an added sustainable impact.

Linking relative and absolute measures

As seen earlier, absolute and relative measures must consider the efficiency and effectiveness of all three levels of the sustainability pillars under a simultaneous character. Also, a strategy for applying these measures is given through questions about the application scenarios. In this sense, it should be asked whether it contributes to sustainability and where resources should be allocated to achieve the most significant possible contribution to sustainability. From this perspective, process managers are always at the forefront of responding to highly essential inquiries in the face of decision making. As a consequence, decision makers must indicate whether the benefit of an investment opportunity exceeds its costs and whether it is more attractive than other investment opportunities. We can learn from making an investment decision that, for this purpose, we must distinguish between direct and opportunity costs. Investments are made only if both cover their direct costs and exceed the benefit that could be achieved if the capital were invested differently (opportunity cost) (Persson & Tinghög, 2020). This same point of view can be attributed to the measurement of contributions to the sustainability of industrial and agricultural segments. Thus, it is recommended that the sustainable measure allows establishing whether the applicability of a resource for a follow-up is sustainable, that is, if the support should be used. Conversely, you must consider where the fund will be allocated for proper use, that is, where the resource reaches the maximum surplus per unit of resource. This decision situation can be expressed in the matrix represented in Fig. 8.3.

To better understand the representation of Fig. 8.3, it is primarily necessary to attend to the questions inserted in the matrix. Therefore, we highlight the relevance whose existing absolute measures should try to answer the first question, that is, If question. In short, this issue is related to determining the quantitative value of the organization that contributes to sustainability. In addition, it is

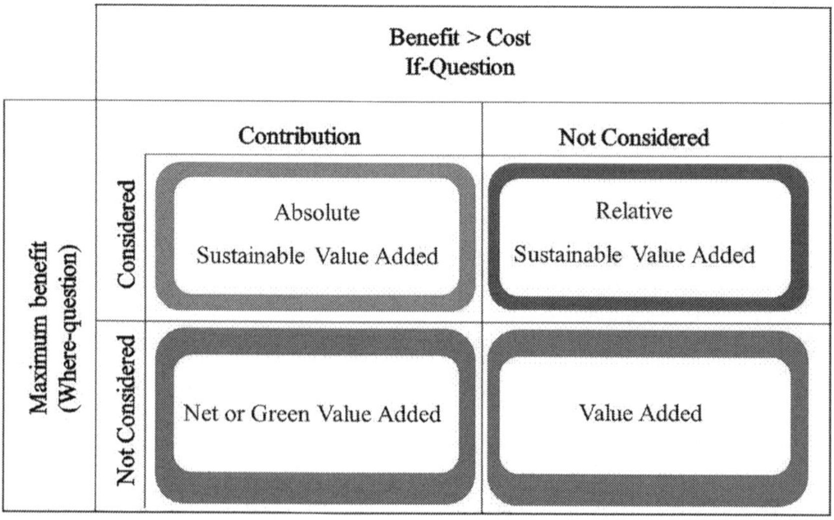

FIG. 8.3 Matrix: if and where. *(Based on Figge & Hahn, 2002.)*

crucial to clarify whether the benefit of using resources will exceed internal and external costs. On the other hand, the differential of this questioning is the identification of the opportunity cost, which is nothing more than the benefit of alternative use of resources, and this, in turn, is not considered in this accounting.

In this sense, the sustainable measure must identify scientific issues such as whether and where environmental and social resources should be applied (see upper left quadrant in Fig. 8.3). However, this measure should cover external costs as well as opportunity costs. In this way, the absolute sustainable added value emerges as a dirty quantitative indicator. Its function is to measure the residual value after the added value of an industrial segment is readjusted to environmental, external, and social costs, in addition to opportunity costs. For more quantitative details, Eq. (8.5) is expressed by:

$$A_{SVA} = VA + EESC + R_{SVA} \tag{8.5}$$

where A_{SVA} is absolute sustainable added value, VA is the value added, $EESC$ is the external environmental, and social cost, and R_{SVA} is the relative sustainable added value.

From the same perspective, once again, this quantitative value attributed to the absolute sustainable added value must contain integral monetary elements. Otherwise, without these references, about the countless external effects caused to the environmental and social performances, it is impossible to generate satisfactory significant results. In this sense, the place where the resources should be allocated, it is possible to respond even if the external costs are unknown. For this purpose, it is necessary to quantify using the relative sustainable added value (see upper right quadrant of Fig. 8.3). Thus, the proposed matrix tends to direct the decision maker of the industrial segment to resolve two distinct issues. Subsequently, the second question, that is, where, will be the focus of attention to the next steps.

In the course of the analysis, both the environmental and social performances of the different segments must be compared with each other to address the local issue. In this way, the corporate contribution related to sustainability is capable of being quantified in full monetary terms. In this context, the relative sustainable added value makes it possible to designate the resources that need to be used to achieve the highest plausible contributions to achieving sustainability. However, although it is not possible to show whether the application of the resource by this segment is sustainable, this indicator can detail how much more viable (that is, in monetary terms) the application of the resource is compared to other segments. Therefore, the relative sustainable added value is defined as the magnitude of the contribution of an industrial segment to obtain more sustainability, through measurement in monetary terms.

In light of what was discussed and presented previously, it is now possible to identify the sustainable added value as an approach to quantify the contributions of sustainability through economic aspects. Thus, it will be necessary to include more in-depth clarifications about the terms of efficiency and effectiveness.

Therefore, in the next topic, we will explain about the derivation of sustainable added value and its application in a unique impact. More precisely, it is considered to keep the argument understandable, whose only eco-efficiency, economy, and eco-efficacy are maintained. That is, to manage the absolute values of economic and environmental performances. In addition, the reduction of eco-efficiency to an environmental impact is included in the process.

Guidelines and contextualization of changes in the concepts of efficiency and effectiveness

Usually, the sectors responsible for the technical-administrative management of industrial and agricultural performance always seek solutions based on eco-efficiency. This, in turn, is basically recognized for directing measures where the consumption of resources and the production of waste are minimized to the environment. Generally, the reduction in resources is related to the decrease in raw materials and energy, aiming at reducing acquisition costs. In short, this measure is characterized by being the pioneer in elucidating the answers that connected economic growth with the protection of the environment.

However, it is known that the term eco-efficiency is not new. Historical estimates project that the introduction of the term was implemented in mid-1992 by the World Business Council for Sustainable Development (WBCSD). Conceptually, eco-efficiency is a procurement strategy based on the premise that it must deliver goods, services, and products at competitive prices, which satisfy human needs and offer the quality of life, progressively reducing the environmental impacts of goods and services, through of the entire life cycle in line with the estimated capacity of the ecosystem to support. In summary, this vision aims to generate products and services whose economic value while reducing the environmental impacts caused in the production chain. Besides, it is essential to note that this link offers a link strengthened by the productivity and profitability associated with social responsibility. Therefore, eco-efficiency is the most efficient use of materials and energy to reduce economic costs and environmental impacts.

To continue to meet the criteria involved in eco-efficiency, it is necessary to follow guiding indicators to remain aligned with its concept. Thus, Fig. 8.4 presents the essential elements of eco-efficiency. From this perspective, considering all these aspects within an industrial segment ends up becoming a complicated practice. For this reason, incentive measures that allow segments to continue using eco-efficiency strategies as a way to protect the environment seem to be a good alternative. However, it is not possible to consider a long-term solution to the environmental problems that challenge humanity. This transformation, in the short term, only results in the systematization of a less destructive process, and therefore, only provides a slowdown in destruction. Thus, reducing environmental impacts creates the illusion of relative short-term improvements, but it is not enough for companies to become truly sustainable. Therefore, more than eco-efficiency, eco-efficacy must be internalized as a guiding principle for corporate sustainability.

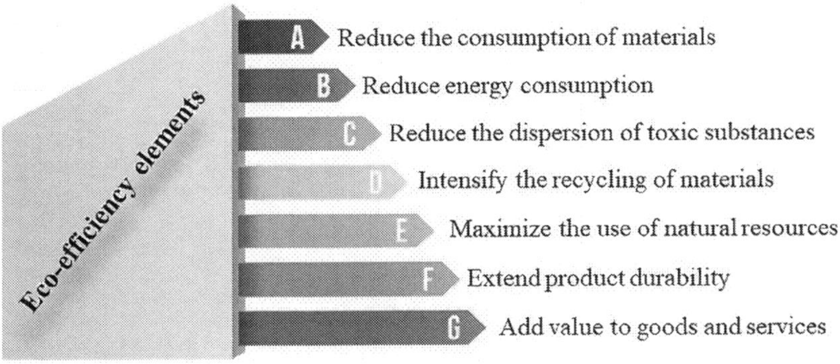

FIG. 8.4 Primary eco-efficiency elements.This figure shows the main elements associated with eco-efficiency.

With the need to be internalized in the industrial segments, efficacy emerged in mid-1998, but without a concrete definition. Subsequently, its concept was attributed to the maxim that carrying out operations correctly without the premise of doing fewer wrong things. This concept is directly based on the definitions of efficient and effective. That is, efficiency encompasses characteristics or qualities or whose equipment creates appropriate or ideal conditions for achieving a particular purpose. On the other hand, effectiveness is known to have the virtue of the power to produce, under normal circumstances, and without needing any other aid, a particular effect to cause an intended result. Given these concepts, at the level of exemplification, it is possible to conclude that eco-efficiency could focus on reducing the energy consumption of artificial light; eco-efficacy would design workplaces that use natural energy. Eco-efficacy is considered the ultimate solution to environmental bottlenecks, as it requires a change in mentality and transformation of business models. Industrial segment managers need to stop damaging the environment and deplete nonrenewable resources, doing more than just postponing environmental destruction. However, to date, eco-efficacy is exceptionally mentioned and is often unknown to most decision makers.

After this brief explanation about the differentiation between eco-efficiency and eco-efficacy from the perspective of economic aspects about sustainability, it is possible to observe that an increase in eco-efficiency can result in a decreasing eco-efficacy. For example, a segment that adopts more eco-efficient measures is subject to demand higher amounts of resources when there is an increase in competitiveness. In other words, improvements in eco-efficiency can be overcompensated by a rebound effect. To further clarify this premise, Fig. 8.5, hypothetically, presents the effects observed in the change in eco-efficiency of a segment in a given period. Besides, characteristics of the absolute change in economic and environmental performances were also considered.

Comprehensively, it is essential to note that this joint assessment allows us to identify the changes that occur between eco-efficiency and eco-efficacy.

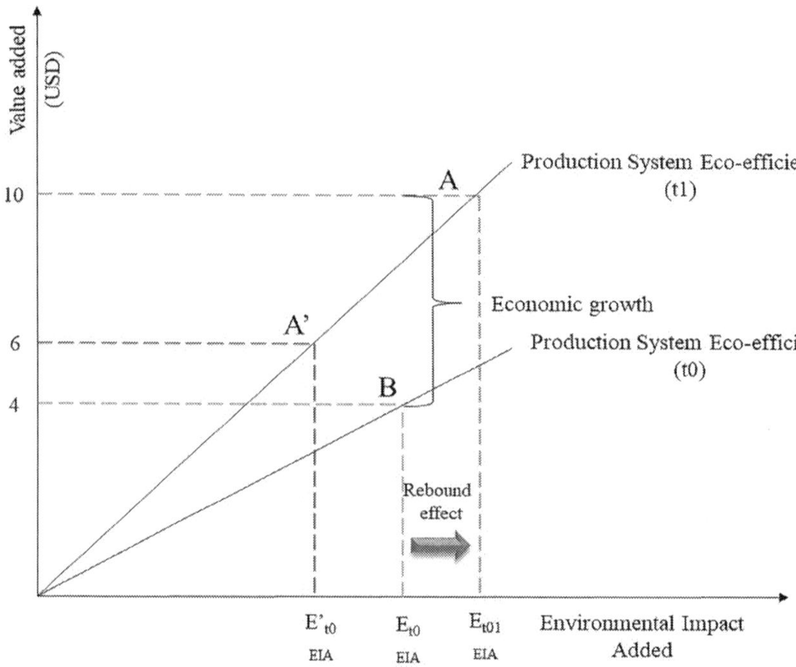

FIG. 8.5 Chart portrayal of the rebound effect for the single impact cases.
It is important to note that the eco-efficiency of the example had a doubled increase between t_0 and t_1. As a result of optimized eco-efficiency, this results in economic growth. However, economic growth is offered by the change in added value between periods. Note that, despite the doubled efficiency, the absolute level of environmental impacts is rebounded. Therefore, it is possible to conclude that economic growth has far outweighed the improvement in eco-efficiency, so that eco-efficiency has decreased. *(Based on Figge & Hahn, 2002.)*

Besides, if there is a reach between the performances of the three pillars of sustainability in at least one level previously reported, emphatically pointed out that a step closer to sustainability could be attributed. As seen in Fig. 8.5, the environmental pillar was impaired, as improvements in economic growth substantially offset the developments employed in eco-efficiency. Therefore, the lower the demand for resources of value created, that is, improvement of eco-efficiency, the more insufficient will be the strategies to compensate for the use of more environmental resources. And, therefore, despite an improved eco-efficiency, the same will suffer more and more aggravating damage.

Opportunity costs: Concepts and application importance

Recently, the search for making individual decisions has become increasingly important in the industrial and agricultural segments. This valuation of decision planning is attributed to the ability to evaluate the scenarios so that all operations occur as expected. For this, there is a managerial consensus that states that making decisions without assessing the global context is undoubtedly quite risky. Thus, estimates generally indicate that the cases in which the actions are successful are minimal, and likely, this strategy will not work in most situations.

In this sense, the concept of opportunity cost emerges as an attempt to express the relationship between scarcity and choice. That is, if no industrial processing stage or activity present in the production system is scarce, the result of all the demands of the other unit operations can be achieved.

In practice, opportunity costs can be assessed directly with cost-effectiveness or cost-utility studies. That is, when two or more interventions are compared, the analysis of the effectiveness of cost utility makes the opportunity cost of alternative uses of resources explicit. Thus, the cost-effectiveness indices, that is, the result of different interventions, allow to compare the opportunity costs of each intervention. However, some decision makers insist on neglecting opportunity costs. Thus, individuals tend to be selective in the use of information, and therefore, focusing on attributes highlighted at the point of decision making.

Additionally, opportunity costs can extend to the three pillars of sustainability. Thus, four main categories of opportunity costs are known (Table 8.3).

Posterior to the conceptualization and the importance of applying and considering opportunity costs, as seen previously, until now, charges were not considered in the calculation. However, to assess the corporate contribution to the opportunity cost of the sustainability of the use of resources must be considered. Thus, for a better understanding of the relationship between sustainable added value and opportunity costs, based on calculation, the opportunity cost can be understood as the lost value. For example, if the levels of economic production in the industrial segment have not been achieved, for reasons such as the reallocation of resources, the benchmark will represent the lost value of the resources required by the segment. Therefore, this lost value is known as opportunity cost. In the financial field, this alternative is generally known as a reference point. From this perspective, we insert opportunity costs into the rationale (Lozano, 2008).

When using the same graphic pattern as mentioned earlier, the costs of opportunities and the eco-efficiency of the benchmark were inserted in Fig. 8.6, which is identified in the image by the dashed EE_b.

Finally, we can now go on to fully substantiate the mathematical model of sustainable added value for scenarios that achieve unique impacts. However, as noted, the improvements that occur during management changes are not always sufficient for substantial sustainability contributions. In this sense, for improved and complete analysis, the inclusion of environmental and social effectiveness is necessary. In the previous example, it can be seen that the improvements related to eco-efficiency were substantially offset by the gradual growth of the economy, resulting in an unwanted degradation of the environmental scenario. On the other hand, it is imperative to take into account the point of view

TABLE 8.3 Exemplification of the different categories of application of opportunity costs.

Category	Definition	Exemplification
Hidden opportunity cost	The hidden opportunity cost is one that, as its name says, is camouflaged. That is, it is not possible to know precisely what will be lost in making a decision. With that, it cannot be measurable.	When choosing between one of the two types of investment, it is easy to know what is the cost involved during an analysis of interest and gains in the deprecated transaction. However, some values are already embedded in certain operations. Thus, they end up being added automatically to a particular type of investment.
Open opportunity cost	Unlike hidden, open opportunity cost is not based on the concept of cloaking. That is, its value is not automatically incorporated in an operation.	For example, in the open opportunity cost, it is necessary to include in the total costs the benefit you lost due to a particular choice made.
Environmental opportunity cost	The environmental opportunity cost is based on the maximum value that can be obtained by enjoying some natural resource.	For example, the burning of a nonrenewable natural resource, such as oil, for the production of fuel ends up preventing its use in other businesses, such as the manufacture of plastic cups. As a result, its use in one industry ended up restricting its use in other areas.
Accounting opportunity cost	This represents the opportunity cost in relation to how much an entrepreneurial segment has had its profit sacrificed by having invested money in one resource in detriment of another.	For example, how much profit was lost (or no longer won) due to the expansion of a new industrial segment. Or even, how much was stopped winning, since a new product line was inserted inside the factory.

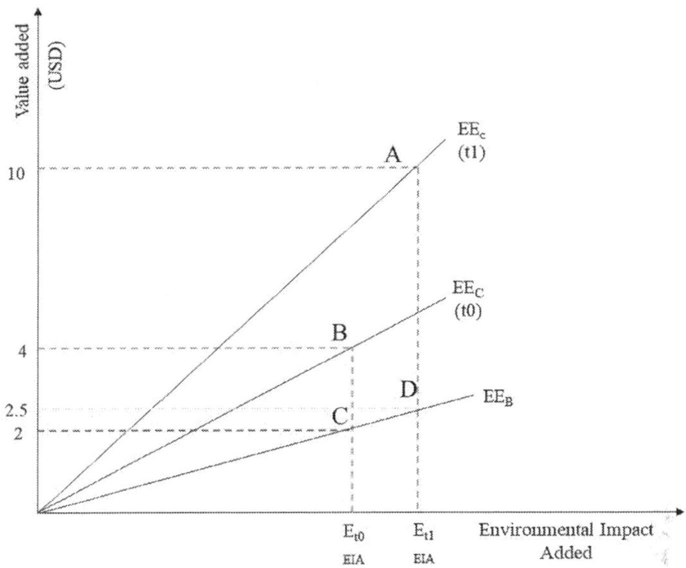

FIG. 8.6 An approach to the deployment of benchmark.
In the example, the choice of the reference parameter depends on the desired significance of the measurement. In the context of corporate contributions to sustainability, the question to be answered is whether a company contributed to the sustainability of a national economy during the period under observation. Therefore, we propose to choose the national economy as a reference. As a consequence, EE_b indicates the eco-efficiency of the national economy, given by the ratio between the gross domestic product (GDP) and the environmental impacts created by all entities of that economy. *(Based on Figge & Hahn, 2002.)*

based on the concept of opportunity. In this way, it is possible to elucidate recurring doubts as to whether the measure was more sustainable since the industrial segment used the extra resources as opposed to the competition.

Therefore, the primary conception of the basis for calculating sustainable added value aims to provide compensation for less efficient users of resources for avoiding environmental impacts. In addition, this concept uses the eco-efficiency of the benchmark to indicate how efficiently the economy appropriates environmental resources. Under this logic, the resources used must be allocated to the industrial segments, which are more eco-efficient than traditional users, so that they become more sustainable.

However, when we recognize that industrial segments use environmental and sometimes social resources to generate value, therefore, other industrials marked as less efficient will make it compatible to create a resource available, that is, put it up for sale, if the same can generate more value then properly use it. Thus, since eco-efficiency presents the cost caused by environmental impact aggregated quantitatively, it also corresponds to the amount that the industry must be able to disburse to convince the competition to waive the resource.

Based on this logic, industrial segments are able to offset the additional costs of environmental impacts caused by the purchase of environmental and social resources from reference industries (Norrman et al., 2020). Therefore, the value of this compensation for environmental or social impact is defined by its respective efficiency of the reference, which we know as the opportunity cost.

Notoriously, compensation does not correspond to the payment of victims of external effects to make them allow these impacts, as is done in comprehensive approaches. Here, the settlement aims to pay the reference companies for having specific resources that they would use. Consequently, the total level of impact is unchanged.

Given this scenario, in order to maximize contributions to sustainability, resources must be allocated to the most efficient industrial segments. However, it is essential to note that industries need to generate additional value, that is, economic growth to purchase extra resources with less efficient sectors. See that the total purchase of resources would result in reduced economic efficiency. Therefore, the purpose of acquiring environmental impacts added by less efficient segments is restricted by economic growth. Should the industry receive more resources and spend more than economic growth to adjust environmental and social compensations, this will cause even more significant damage in terms of economic effectiveness.

In this sense, to identify quantitatively whether the industry has sufficient economic funds to pay for the reduction of the additional environmental impact it has generated, it soon becomes necessary to compare the available additional funds. This comparison is usually given by economic growth with the cost and amount of other environmental impacts. This modeling is given through Eq. (8.6):

$$SusVA_{Si} = EG - EE_b \times (EIA_{t_1} - EIA_{t_0}) \tag{8.6}$$

where VA is the being value added, EE_b corresponds to eco-efficiency, EIA is the environmental impact added, and t_0 and t_1 indicating periods. Note that price is a factor corresponding to the benchmark eco-efficiency. In addition, the amount is equivalent to the additional environmental impacts caused. In this way, the manufacturing company can pay compensation for impaired environmental performance provided that economic growth is at least equal to the cost of compensation.

Subsequently, for cases of single impacts, a sustainable added value is created if an additional value has been created after compensation for the negative changes in eco-efficacy. Sustainable value added for the case of a single impact, therefore, it is equal to economic growth minus the cost of compensation. This is expressed by Eq. (8.7):

$$SusVA_{Si} = EG - EE_b \times (EIA_{t_1} - EIA_{t_0}) \tag{8.7}$$

where $SusVA_{si}$ is the sustainable added value for the single impact case, EE_b is the eco-efficiency of the benchmark, EG is the representation of economic growth ($VA_{t_1} - VA_{t_0}$), and $EIA_{t_1} - EIA_{t_0}$ is the additional environmental impacts that need to be compensated.

On the other hand, the sustainable added value aiming at analyzes of cases of unique impacts investigates the changes in the economic, environmental, or social performance of an industrial segment during a determined period whose contribution to sustainability concerning a reference. This indicator is expressed in monetary terms by how much the performance of an industry in a known period ($t_0 - t_1$) has contributed to making the national economy more sustainable. For this purpose, an opportunity cost perspective is presented. On the other hand, *SusVA* analyzes a situation of a constant general level of consumption of resources. The latter, in turn, calculates and expresses in monetary terms the extra value that industry creates under the condition of a constant general level of resource consumption.

In parallel, the multiple impact assessment scenarios consider social impacts. Thus, to assess social efficiency and effectiveness, social performance is established as the social harms, that is, accidents at work. As a consequence, efficiency and social effectiveness are operated in a similar way to eco-efficiency and efficacy. In this way, all eco and social efficiencies are related, of course, to a common numerator, that is, the added value. In the case of multiple impacts, certain environmental damages and social consequences represent the complete set of resources that are used by an industrial segment to generate a certain level of economic production. Therefore, any environmental or social impact of the production process or reference is characterized by eco-efficiency (EE_i) or specific social efficiency (SE_j) and eco-effectiveness (EIA_i) and social effectiveness (SIA_j). Notably, efficiency and effectiveness are always related to a specific period, so in the case of multiple impacts, economic and social efficiencies provide additional information. Thus, the ratio of one efficiency to another shows how much more one impact contributes to the VA than the other implications. This makes it possible to measure the combined effect of a mixture of environmental and social impacts on the creation of value.

Given this context, to establish sustainable added value in multiple impacts, it is possible to analyze the reasoning analogous to the previous example. Thus, when economic growth exceeds the cost of compensation for any negative changes in eco-efficacy, the sustainable added value for a single impact is positive. On the other hand, in the case of multiple impacts, as all environmental and social impacts are considered, economic growth must compensate for changes in a mix of environmental and social effects. Therefore, a positive sustainable added value for multiple impacts occurs as long as economic growth exceeds the joint contributions to the creation of value from negative changes in effectiveness. More precisely, the cost of compensation for all additional environmental (AIA_i) and social (SIA_j) impacts are summarized and compared to economic growth. In view of the earlier, Eq. (8.8) presents the appropriate modeling to quantify sustainable added value in multiple impacts:

$$VA_{t_1} - VA_{t_0} \geq 1n + m \sum i = 1n \times EE_{i,b} \times (EIA_{i,t_1} - EIA_{i,t_0}) \\ + \sum j = 1m \times SE_{j,b} \times (SIA_{j,t_1} - SIA_{j,t_0}) \quad (8.8)$$

where VA_{t_1} and VA_{t_0} are the added values of production system at t_1 and t_0, n and m are the numbers of relevant environmental and social impacts. EIA_{i,t_0} and EIA_{i,t_1} represent eco-effectiveness for environmental impact i in t_0 and t_1, and SIA_{j,t_0} and SIA_{j,t_1} social effectiveness for social impact j and m t_0 and t_1. $EE_{i,b}$ and $SE_{j,b}$ are the economic or social efficiency of the benchmark for the environmental resource i, i.e., the social resource j, respectively.

In general terms, the right side of Eq. (8.8) describes the sum of all costs and revenues from changes in any social or economic effectiveness of the industrial segment, respectively. As mentioned earlier, the ratio of two efficiencies shows how much more an impact contributes to value added than another impact. Therefore, efficiencies are used to weigh changes in effectiveness. Thus, by dividing by the number of impacts, that is, the term $n + m$, we obtain the weighted average and, therefore, the sum of all costs and revenues. So, the sustainable added value for multiple impacts is given through Eq. (8.9):

$$SusVA = EG - 1n + m \sum i = 1n \times EE_{i,b} \times (EIA_{i,t_1} - EIA_{i,t_0}) \\ + \sum j = 1m \times SE_{j,b} \times (SIA_{j,t_1} - SIA_{j,t_0}) \quad (8.9)$$

where $SusVA$ is the sustainable added value for multiple impacts, EG is the representation of economic growth ($VA_{t1} - VA_{t0}$), VA_{t1} and VA_{t0} are the production systems added value in t_1 and t_0, n and m are the numbers of relevant environmental and social impacts, $EIA_{i,t0}$ and $EIA_{i,t1}$ represent eco-effectiveness for environmental impact i in t_0 and t_1, and $SIA_{j,t0}$ and $SIA_{j,t1}$ social effectiveness for social impact j and m t_0 and t_1. $EE_{i,b}$ and $SE_{j,b}$ are the economic or social efficiency of the benchmark for the environmental resource i, i.e., the social resource j, respectively.

However, it is worth noting that a positive $SusVA$ indicates that an industrial segment managed to generate an extra value compared to a reference, maintaining the general consumption of each resource used at the level of the previous period. In contrast to compensating victims for accepting externalities, environmental and social resources are reallocated among different users. In this way, money transfers are only used to get affected users to allow reallocation. Thus, the sum of costs and revenues for which economic growth is adjusted reflects the net amount of money resulting after the liquidation of all changes in the socioeconomic and corporate effectiveness of the reference companies. In terms of exemplification, this is equivalent to the company's net contribution to the economic growth of the benchmark, often measured by GDP in the case of an entire economy.

Finally, regardless of the perspective adopted, it is essential to emphasize that only the industrial and agricultural segments whose efforts are required to improve economic, environmental, and simultaneously social efficiency will, in fact, contribute to sustainability. However, as noted during the topic, it assumes that social and environmental resources are used by industries only to generate economic value and, therefore, institutions agree to refrain from using a resource, if compensated for the lost value.

References and recommended reading

Papers of particular interest have been highlighted as:
• of special interest
•• of outstanding interest

Baumgartner, R. J. (2009). Organizational culture and leadership: Preconditions for the development of a sustainable corporation. *Sustainable Development, 17*(2), 102–113.

Baumgartner, R. J. (2014). Managing corporate sustainability and CSR: A conceptual framework combining values, strategies and instruments contributing to sustainable development. *Corporate Social Responsibility and Environmental Management, 21*(5), 258–271.

Baumgartner, R. J., & Ebner, D. (2010). Corporate sustainability strategies: Sustainability profiles and maturity levels. *Sustainable Development, 18*(2), 76–89. https://doi.org/10.1002/sd.447.

Baumgartner, R. J., & Rauter, R. (2017). Strategic perspectives of corporate sustainability management to develop a sustainable organization. *Journal of Cleaner Production, 140*, 81–92.

Blühdorn, I. (2017). Post-capitalism, post-growth, post-consumerism? Eco-political hopes beyond sustainability. *Global Discourse, 7*(1), 42–61.

Dabbous, A., & Tarhini, A. (2019). Assessing the impact of knowledge and perceived economic benefits on sustainable consumption through the sharing economy: A sociotechnical approach. *Technological Forecasting and Social Change, 149*, 119775.

Doane, D., & MacGillivray, A. (2001). Economic sustainability: The business of staying in business. *New Economics Foundation, 1*, 1–52.

Figge, F., & Hahn, T. (2002). *Sustainable value added—Measuring corporate sustainable performance beyond eco-efficiency*. Lüneburg: Centre for Sustainability Management.

•• Figge, F., & Hahn, T. (2012). Is green and profitable sustainable? Assessing the trade-off between economic and environmental aspects. *International Journal of Production Economics, 140*(1), 92–102.

Jänicke, M. (2012). "Green growth": From a growing eco-industry to economic sustainability. *Energy Policy, 48*, 13–21. https://doi.org/10.1016/j.enpol.2012.04.045.

• Lozano, R. (2008). Envisioning sustainability three-dimensionally. *Journal of Cleaner Production, 16*(17), 1838–1846. https://doi.org/10.1016/j.jclepro.2008.02.008.

Norrman, J., Söderqvist, T., Volchko, Y., Back, P. E., Bohgard, D., Ringshagen, E., … Rosén, L. (2020). Enriching social and economic aspects in sustainability assessments of remediation strategies—Methods and implementation. *Science of the Total Environment, 707*, 136021.

• Persson, E., & Tinghög, G. (2020). Opportunity cost neglect in public policy. *Journal of Economic Behavior & Organization, 170*, 301–312.

Peters, M. S., Timmerhaus, K. D., & West, R. E. (2003). *Vol. 4. Plant design and economics for chemical engineers*. New York: McGraw-Hill.

Santoyo-Castelazo, E., & Azapagic, A. (2014). Sustainability assessment of energy systems: Integrating environmental, economic and social aspects. *Journal of Cleaner Production*, *80*, 119–138.

Schneider, A. (2015). Reflexivity in sustainability accounting and management: Transcending thee-conomic focus of corporate sustainability. *Journal of Business Ethics*, *127*(3), 525–536. https://doi.org/10.1007/s10551-014-2058-2.

Tharp, J., & Chadhury, P. D. (2008). *Corporate social responsibility: What it means for the project manager*. Paper presented at PMI® Global Congress 2008—North America, Denver, CO Newtown Square, PA: Project Management Institute.

Chapter 9

Sustainability metrics and indicator strategies for a modern society

As environmental awareness has progressed due to external pressure, both through legal and societal aspects, many industrial manufacturing segments have recently adopted operational strategies to reduce their environmental impacts through the practice of process efficiency improvements and waste minimization. As a result, numerous environmental, economic, and social benefits are expected to be consolidated.

Although the term sustainability can be conceptualized in several ways, its latent premise consists of the no-dissociation of the economic indicator against environmental security and the well-being of world society. Given this scenario, sustainability metrics and indicators are developed to establish a robust relationship between environmental, economic, and social indicators within the production process, making it an excellent link for industrial and agricultural segments to begin incorporating the sustainability objective into making-decisions.

In recent years, as seen in previous chapters, countless assessment methods and tools for environmental and sustainability performance have been developed. Most of the tools discussed present alternative metrics for implementing and meeting basic criteria such as making standard process assessments useful for managerial and expressive service applications, reproducible for consistent and comparable results, and making cost benefits feasible.

The fundamental structure of sustainable metrics comprises five basic indicators, namely, material and energy intensity, water resource demand, pollutant, and toxic emissions. However, secondary metrics may be supplemented within these major categories established in accordance with the accuracy required in other areas of expertise. Therefore, the origin of the various impact categories assessed by the life cycle assessment tool emerges as complementary metrics to provide greater decision support by combining a larger number of components to be assessed. In contrast, some experts suggest that determinations that regulate multiple components are in fact inflexible and inadequate to make comparisons between products and processes. Thus, they propose that it is more

Sustainability Metrics and Indicators of Environmental Impact. https://doi.org/10.1016/B978-0-12-823411-2.00004-9

advantageous to have a basic set of simple and widely applicable applications and to focus complementary metrics to meet the needs of specialized industries.

Based on Maslow's theory, a new conception of sustainable metrics is proposed in order to adapt a pyramid of needs, involving the main environmental tools. In this context, the foundation of the pyramid is the application of the thought life cycle assessment a tool, which suggests that there is a recurring need to become aware of how everyday activities affect our ecosystems. Then the single emission method, associated with carbon emissions and followed by the full environmental burden assessment tool, is suggested by life cycle assessment. Next, economic strategies related to life cycle cost analysis, as well as carbon market transactions, establish the feasibility of applying the processes. And, finally, a sustainable life cycle assessment complements the direct linkage of the three main indicators of sustainability metrics (Fig. 9.1).

However, when viewed as a pyramid, the difficulty of balancing the three pillars of sustainable development is more of a sequencing guideline to follow than, indeed, a goal to be achieved. Yet, building on crucial assumptions

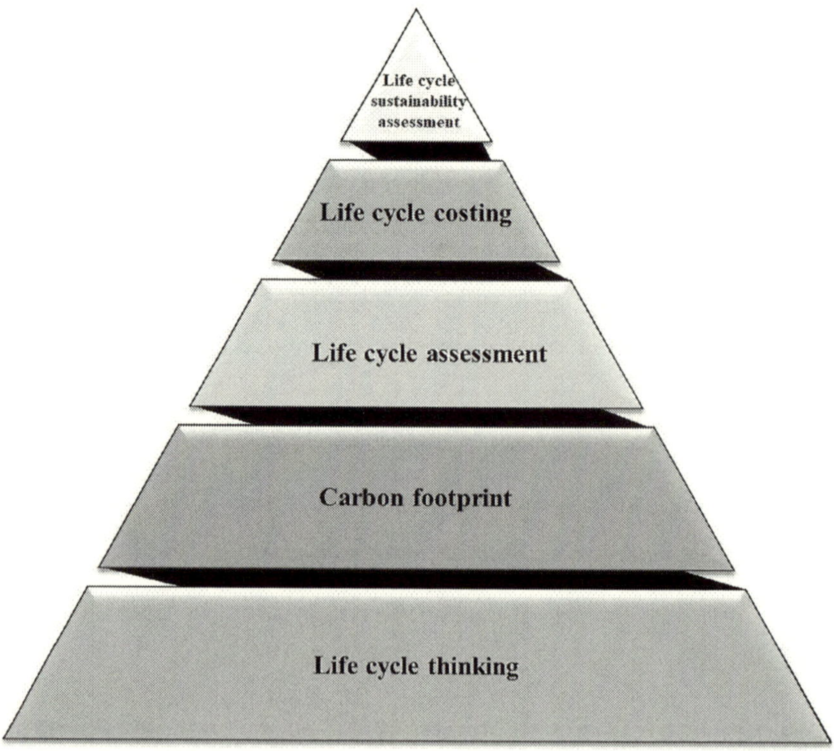

FIG. 9.1 Pyramid for life cycle based environmental and sustainability assessment approaches. This figure represents the hierarchy of consistent application of the life cycle assessment and its assistive tools.

published in previous research reports, the main predetermined sustainability metrics do not legitimately portray the three aspects of sustainability and present complex application methodologies. Therefore, the use of joint indicators enables simple parity between processes and products. However, the loss of data captured by the analysis investigation is not pertinent to the adherence of specific improvement measures. Thus, the use of quantifiable indicators promotes the benefit of assessing the process changes or policies required to make processes and products more sustainable.

Theoretically, Sikdar (2009) proposed a structural model to investigate the delineation of the environmental burden of industrial systems. This structure was based on the basic principles of choosing environmental indicators and how independence requirements would facilitate changes in the concepts of some indicators or how they are quantified, according to the attributes and types of data available without affecting others. Thus, Fig. 9.2 considers the three

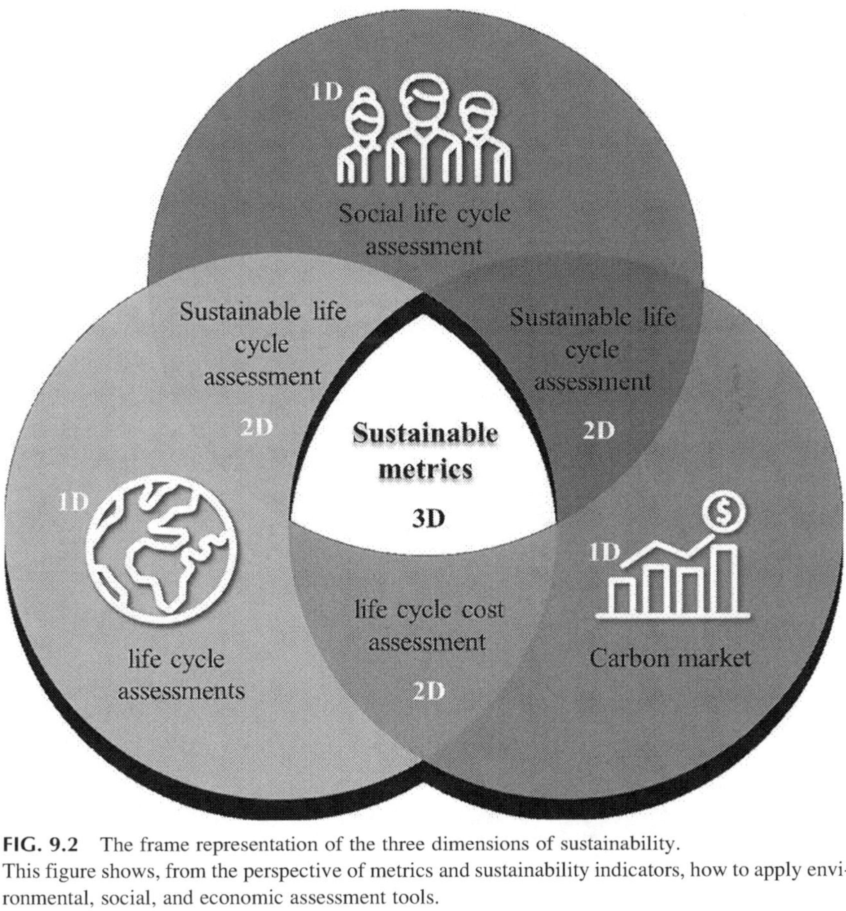

FIG. 9.2 The frame representation of the three dimensions of sustainability.
This figure shows, from the perspective of metrics and sustainability indicators, how to apply environmental, social, and economic assessment tools.

sustainable dimensions represented by hierarchical groups, where one-dimensional (1D) indicators provide information about only one dimension of sustainability: economic, ecological, or societal; two-dimensional (2D) indicators provide information simultaneously about two dimensions of sustainability: socio-ecological, socio-economic, or economic-ecological; and three-dimensional (3D) indicators or sustainability indicators provide information about all three dimensions of sustainability.

Notably, to obtain sustainable metrics you must first quantify the metrics in 3D. For a broader understanding, in some situations when analyzing 3D metrics, processes may have a system improvement as a result of certain changes made to process design, in which case it may be possible to conclude that the optimized process is more sustainable. On the other hand, since the results are ambiguous, one must examine the 2D metrics, and then the 1D metrics for decision-making. This organizational methodology does not indicate which impact categories should be used for any specific situations. Therefore, the need to correctly apply the choice of impact categories to be evaluated will still depend on the analyst. At the level of exemplification, the number of impact categories and the metric level achieved for a given industrial process will be heterogeneous with other manufacturing processes that may require many unit operations, as well as other operating requirements.

Similarly, for a process understudy to be quantified according to 3D metrics, the organization, as well as the definition of the categories, must be thoroughly established. The economic, social, and environmental burden analysis tools already seen can be used to determine these metrics. Thus, to quantify 1D metrics, we can use the tools of life cycle assessments (environmental) and carbon markets (economy), while the social life cycle assessment (social) can be used to determine aspects related to product production, waste treatment costs, as well as employment taxes. On the other hand, tools such as life cycle assessment (environmental and social), life cycle cost assessment (environmental and economic), and sustainable life cycle assessment (economic and social) can be considered as assistants of 2D metrics. In this sense, the 3D metrics encompass all other social, environmental, and economic analyzes in order to actually consider the total sustainable metrics of processes and products.

Given this context, the determination of the sustainable metrics of a process is generally proposed for the analysis of real industrial processes and agricultural systems that require more robust investigations, resulting in more intensive procedures compared to the hierarchical evaluation structure. Therefore, when choosing sustainable metrics, it is suggested that the processes are directly associated with the sustainability tripod, without tendencies for the other impacts (Sikdar, Sengupta, & Mukherjee, 2017).

The logic of sustainability metrics follows the same reasoning as impact categories, that is, the smaller the metric, the more efficient the process will be. Thus, the guideline that smaller is better can be used to assess the relative performance of manufacturing processes in terms of impact per production unit.

In addition, for many systems, fuel energy consumption is the most useful energy intensity metric for benchmarking performance and tracking progress. These metric captures process improvements such as improved heat recovery, better heat integration, or increased throughput. This also reflects improved power and steam generation efficiency through technologies such as cogeneration.

However, there are some too frequent circumstances about decision-making that appear to be obstacles to quantifying sustainable metrics. These, in turn, need to be circumvented with proper procedures to ensure the reliability of the results. Therefore, trade-offs between the three pillars of sustainability must be approached with the utmost caution to maintain sustainable balance. In addition, another crucial factor in the following schemes still allows this to be done in a transparent rather than implicit manner if decision-makers decide to apply quantitative weighting. The utility analysis approaches chosen to have the advantage that, in addition to quantitative criteria, qualitative assessments can also be considered and presented in the same assessment scheme.

Sustainability tripod dimensions

Environmental dimension

The key assessment tool designed by the environmental dimension is based on life cycle assessment, representing through impact categories a fundamental quantitative instrument to guide possible damage to the environmental sphere. However, this tool, while necessary, is not fully sufficient to track progress and to establish long-term strategies on sustainability metrics. There are numerous obstacles within the mapping of system boundaries that result in processes that are difficult to quantify or that are more qualitative in nature that need to be considered in all decision-making processes, and consequently in the final elaboration of the main global policies for management of productive systems.

In addition, another strategy for addressing environmental issues has emerged with the creation of natural resource protection management programs. However, some obstacles still stand in the way of trying to bring about the environmental excesses of other countries in the search for commercial guarantees. Thus, greater importance is given to the environmental management required exclusively in view of the substantial development of the ecological crisis.

Economic dimension

Many traditional cost accounting systems lead to incorrect investment decisions regarding environmental costs. One of the problems of quantifying the economic dimension is determining the demolition and recycling costs that appear outside the bounds of the traditional accounting system. Moreover, historically, economics tools have application restrictions in an environmental scenario, resulting in certain barriers being the uncertainties of the decision-making team. From an economic point of view, assessment tools are not considered as alternatives capable of preserving ecosystems and may make irreversible changes in environmental management processes. Therefore, this dimension is widely known for ignoring the natural environment, as well as simplifying environmental problems as it is presented exclusively as monetary figures.

Against this background, a popular way to solve this problem has been to propose the use of life cycle cost. However, the life cycle costing tool can be applied in two ways, being the sum of the total costs of a process or product, but disregarding the long period of its useful life. On the other hand, this tool allows comparative cost assessments to be performed over a specified period of time, taking into account all relevant economic factors, both in terms of initial costs and future operating costs.

Another recently proposed alternative tool is known as the life cycle cost assessment, which, still in an immature stage, presents itself as the solution to the bottlenecks of evaluating the systemic process of the life cycle costs of a process or product. Its main objective is to minimize the disadvantages of traditional economic tools and maximize the advantages of environmental analysis tools by providing identification of environmental consequences and assigning monetary value measures to those consequences.

Social dimension

The social dimension of sustainability captures the impact of an organization, product, or process on society. The main social objective of applying tools with environmental indicators is to quantify social values and goals in different sociopolitical environments. However, like the economic dimension, the social dimension is presented in a slightly more complex way, since social measures do not occur easily.

Traditionally, the selection of criteria about social issues and their quantification is one of the main challenges in implementing the concept of sustainability. While much has been advanced in these parameters, there are still latent needs for research and consensus from the stakeholders involved. Currently, as seen in Chapter 5, there is no uniform use of a standard set of indicators, but operationally applicable indicators are available. Extensive research on sustainable indicators focusing on the social dimension has resulted in numerous proposals for different social goals and indicators, yet social life cycle assessment is still in its infancy.

In this regard, the next step in addressing the great challenge of establishing secure sustainable metrics is to establish a collection of indicators solely for public consultation through a consistent and frequently updated database. However, this will result in a substantial effort encompassing environmental lifecycle approaches, as well as economic approaches.

Moreover, once these criteria are established, other bottlenecks of technical inconsistencies emerge as misapplications of application between environmental and social tools. This is because both life cycle analysis and social life cycle analysis are based on the designer thought analysis tool. However, although they are similar in many respects, these tools generate inconsistencies as one of them refers to subjective factors as they are very difficult to quantify and their impact changes with business management behavior even if within a business. In the same locality, there may be significant differences in this respect at the company or sector level, as opposed to the well-established environmental indicators.

Another recurring factor in the social dimension is the likelihood of double counting, for example, economic and social impacts. This is because, unlike environmental life cycle analysis, where impacts are generally mostly negative, environmental impacts can be noticeably positive. However, by applying the notion of externality, this barrier can be avoided by considering as economic impacts all those effects that are adequately reflected in prices and associated quantities, while social impacts are those that are not adequately captured by prices, i.e., where prices are also high or very low. The problem with this recommendation is that it may sound convincing on paper, but it faces a number of practical problems, including the economically correct price. In short, in order to propose consistent tools for quantifying sustainable metrics, it is necessary to adopt a range of holistic environmental, economic, and social analytics tools to use them against the sustainability tripod (Fig. 9.3). Therefore, the overarching idea would be a three-dimensional scheme that will consist of a general assessment with process-specific indicators in order to result in a real application utility approach.

Finally, we can say that sustainability metrics and specific indicators are useful tools. Its usefulness in making project choices and policy decisions remains limited. Consequently, an exploration of overlapping concepts related to preference index is suggested. In this way, making project choices and making policy decisions can provide a broader framework and help address multiple competing sustainability-related goals and priorities.

Suggestions for the development of sustainability metrics and indicators

Conceptually, it is possible to define an indicator as an operational representation of an attribute, that is, the variation of quality criteria, characteristics, as well as properties of a given system, and is established by quantitative or qualitative variables (for example, numbers, graphs, colors, symbols, or function of variables). Besides, this indicator may include its value associated with a reference value.

Not unlike, sustainability indicators should reflect the reproducibility of the way a given society uses its environment through quantifiable variables that

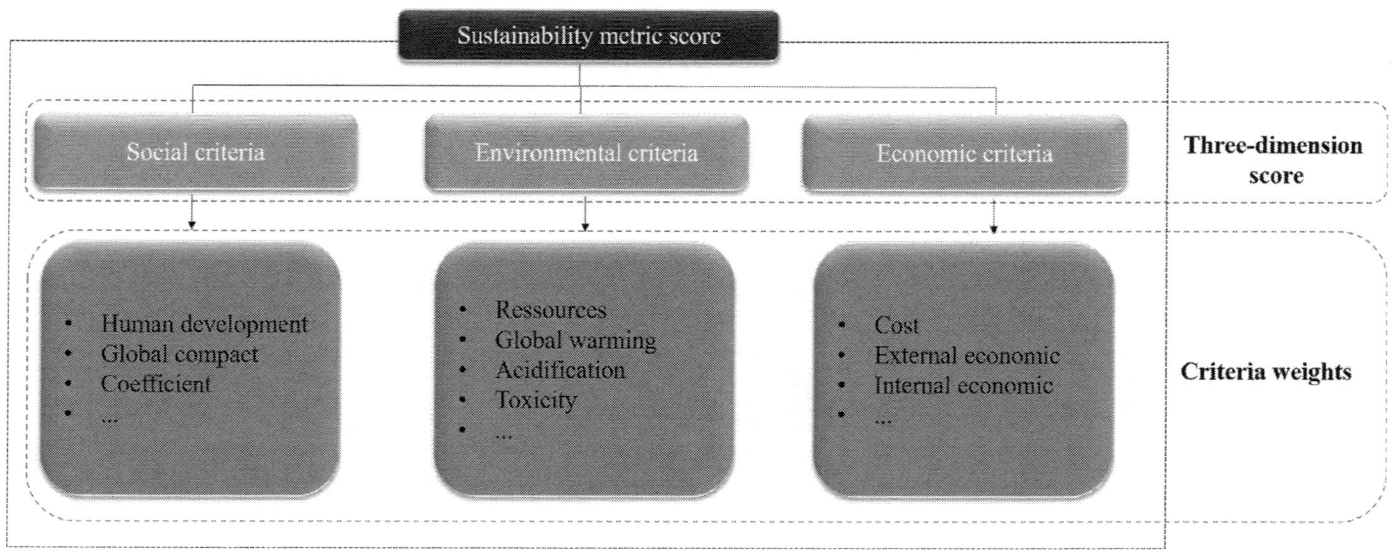

FIG. 9.3 Global assessment steps for sustainability metrics.
In summary, the figure shows the three pillars of sustainability to obtain the metrics score.

express the closest to qualitative damage. From this perspective, sustainability indicators do not merely reflect environmental conditions or pressures on the environment. Still, they indicate the degree to which specific environmental influences or impacts the land can deal with in a long-term perspective, without being affected in its structures and underlying processes.

To achieve this level of assessment, it is necessary that monitoring progress toward sustainable development requires, first of all, the identification of operational indicators that provide manageable units of information on economic, environmental, and social conditions. However, it is suggested that the establishment of normative indicators be built considering the real developments of the objectives to be evaluated.

On the other hand, notably, the current deficiencies have led to the need for a widely shared, comprehensive, and sufficiently detailed methodology to assess consistent environmental indicators. In this sense, to establish such a method, it is essential to first clarify the critical issues in life cycle assessment in a comprehensive manner and to review the various approaches that have been developed to deal with these issues.

Indeed, as seen throughout the chapters, the variety of sustainability indicators poses a significant problem, especially as policymakers require an aggregated index that can be unambiguously interpreted and easily communicated to the general public. Thus, we found that, although most sustainability indices are considered concise and transparent, they fail to meet fundamental scientific requirements (Böhringer & Jochem, 2007).

As suggested in the previous chapters, we recommend the use of more diversified sustainability indicators in future assessments of production systems to expand the limits of the system and improve the totality. In addition, we suggest that when choosing sustainability indicators, it is crucial that they meet some quality criteria to ensure that the indicators can adequately reflect the sustainability status and changes in systems (Agamuthu & Hotta, 2014). Thus, we can gather four main criteria: the representativeness, comparability, responsiveness, and scale of the indicator.

Briefly, the representativeness suggests that the indicator should clearly represent the property of the sustainability aspect of the measured system. However, comparability aims to find and use the most tangible, quantifiable, and comparable attribute of an aspect of sustainability when deciding on an indicator. On the other hand, the responsiveness must accurately reflect the status and changes in the aspect of sustainability that it represents. And, finally, the scale of the indicator should pay attention to the scale of an indicator, for example, does it represent a system-wide problem or a specific problem during a specific phase of the food systems? If an indicator is on a smaller scale, what could be its comprehensive indicator on a larger scale, and what interconnections does it have with other indicators on a similar scale?

In addition, basic criteria of interpretation must be observed on how sustainability should be understood considering its organizational principles applied in a given socio-environmental context. Likewise, the structuring of information can be understood as the multidimensional complexity inherent to sustainability and must be structured in operational information units (for example, indicators) and properly communicated to feed the decision-making process. Besides, the influence of sustainability information must exert real influence on decision-making and the effective implementation of sustainable development.

Finally, regardless of the strategy for creating sustainability indicators, or how the environmental analysis tool will be applied, it is important to be aware that there is no negative impact on human activity. Even a replacement of natural resources (such as replanting of plant species or protected reproduction of animal species is only a replacement, as per the definition itself). Productive, human industrial activity, even the occupation of areas for any purpose, is always an impact and can only be minimized or compensated (for example, by replenishing but not necessarily in the same ecological niche/physical space—a desert not it is a forest, and a ravine has not undoubtedly had a forest in the last millennia). Besides, some formerly existing species can be forgotten, excluded, unknown in the assessment, or already extinct and, therefore, irrecoverable.

Beyond sustainability

Unquestionably, the introduction and propagation of sustainability concepts had a profound impact on the broad and complex discourse that involves industrial and agricultural segments, as well as the relationship with the environmental crisis and social development. However, these concepts, beyond promoting the reconciliation of the three main pillars of sustainability, basically require a balance between the many processes and operations that are inserted in the production system over a long period.

It is evident that decision-making in the face of sustainability seems to require substantial reductions in anthropological activities. Among these actions, the reduction of natural resources and the improvement of working conditions in the social sphere is indisputable. However, to date, there is no evidence that these measures are, in fact, sufficient to achieve the sustainability seal.

Besides, although it is attractive for industries to have this authentication, in practice, generally, the measures to be taken are uninspiring and quite inaccessible. Contradictions suggest that sustainability is more attainable when factors and parameters involving lower levels are considered. In contrast, some authors and decision-makers indicate that sustainability issues can only yield relevant results, at levels of larger systems, where there are voluminous and complex operational interactions.

Indeed, since the delineation of the term sustainability, many theories have been generated regarding the veracity of the criteria for considering a sustainable process. However, many problems remain unsolved. Thus, the most widespread theory among managers indicates remuneration criteria. In short, environmental performance is directly related to the capacity for economic growth. That is, if economic growth is equal to or higher than the cost of compensation for deteriorated environmental performance, this process may be considered sustainable.

In view of this argument, sustainability, as it is currently adopted in many of its designs, is considered mainly an economic and anthropocentric notion. Thus, there is no doubt that there is a necessary moral restructuring, whose integration includes issues related to the extraction and conscious preservation of natural resources. Only then will it be possible to develop a more precise understanding of sustainability.

Therefore, it is crucial to discern points of view. In short, sustainability is based on its technical-scientific character. On the other hand, the ethical notion is necessary for us to test sustainable development. Thus, according to Kothari (1994), the technical and scientific idea of sustainable development does not reach the heart of the environmental crisis. A new approach to sustainable development would be needed, which would require an ethical shift in thinking. More precisely, a change in values in such a way that nature is valued in itself and for its life support functions, not just because it can be converted into resources and commodities to feed the engine of economic growth.

However, it is known that because of the central sustainability theories, return capital presupposes natural capital. Therefore, the most critical point for the consolidation of any process is the economy. Thus, as previously seen (see Chapter 8), the generation of added value can be the most plausible tool for the strengthening of sustainable processes.

On the other hand, the generation of value attributed to products and industrial processes may affect the value of products to end consumers. Given the environmental scenario of the processes, emerging issues are highlighted, such as the dilemma granted to sustainable products. In fact, are the most expensive products really the most sustainable? Also, are consumers willing to pay the high price related to environmental quality?

In fact, from the acquisition of raw materials to the shipment of the final product, almost all stages of the manufacturing process and production of sustainable products cost more than traditional products. Consequently, this trend reflects consumer behavior when buying sustainable products at a higher price. Therefore, to extract consumers' willingness to pay for environmental quality, companies rely on eco-labeling programs to verify that their products comply with the standard set by government incentive programs. However, most consumers are not willing to pay more or invest in sustainable products to obtain long-term environmental benefits. Thus, an effective measure is based on supply and demand models. Therefore, only an increase in demand can result in a reduction in the price of the product.

Once again, Kothari's theory emerges and becomes essential. Consumers must understand the power of their purchasing decisions when it comes to caring for the environment and choose brands and companies whose actions align with their values and expectations beyond the purchase of a product, even if they cost a little more. In this way, as new industrial and agricultural process systems are planned and implemented, sustainability analyses are conducted to address potential environmental impacts through the life cycle analysis tool. Besides, understanding the policy directives associated with production processes is essential for sustainable tool implementers to explore in-depth information that would complement the data for certain categories. This, in the medium term, will allow for major information and sustainable assessment with decisions that are not readily apparent or externally accessible.

Finally, the different approaches from the sustainable tools, in the last 30 years, have shown remarkable advances and have been recognized as mathematical tools of wide application scope. Thus, in the current state, despite the numerous application bottlenecks, life cycle assessment, life cycle costing, social life cycle assessment, and life cycle sustainability assessment are moving toward application success in the assessment of sustainability metrics and indicators. Therefore, articulating these visions by integrating each pillar of sustainable development is not only a task for researchers, decision-makers, and administrative sectors, but it also requires the efforts of all people in the spheres of life. Under this circumstance, there is much to be learned from implementing sustainability. However, until we reach an agreement on the importance of producing a positive outlook, the best we can hope for is that we will be confused over and over again with successful economic growth. In these times, the challenge of getting in the way is important and should not be taken lightly. But, a stop on the way should not be confused with the end of the sustainable journey.

References and recommended reading

Agamuthu, P., & Hotta, Y. (2014). Indicators as a tool to evaluate waste management efficiency. *Waste Management & Research*, *32*(12), 1147–1148.

Böhringer, C., & Jochem, P. E. (2007). Measuring the immeasurable—A survey of sustainability indices. *Ecological Economics*, *63*(1), 1–8.

Kothari, R. (1994). Environment, technology and ethics. In L. Gruen, & D. Jamieson (Eds.), *Reflecting on nature: Readings in environmental philosophy* (pp. 228–237). New York: Wiley-Blackwell.

Sikdar, S. K. (2009). On aggregating multiple indicators into a single metric for sustainability. *Clean Technologies and Environmental Policy*, *11*(2), 157–161.

Sikdar, S. K., Sengupta, D., & Mukherjee, R. (2017). Systems, indicators, and sustainability assessment. In *Measuring progress toward sustainability* (pp. 59–91). Cham: Springer.

Annex

Annex 1: Table A.1.

TABLE A.1 Global warming potentials (GWP) relative to carbon dioxide for the 100-year time horizon.

Industrial designation or common name	Chemical formula	GWP for 100-year time horizon (at date of publication)
Carbon dioxide	CO_2	1
Methane	CH_4	25
Nitrous oxide	N_2O	298
Substances controlled by the Montreal Protocol		
CFC-11	CCl_3F	4750
CFC-12	CCl_2F_2	10,900
CFC-13	$CClF_3$	14,400
CFC-113	CCl_2FCClF_2	6130
CFC-114	$CClF_2CClF_2$	10,000
CFC-115	$CClF_2CF_3$	7370
Halon-1301	$CBrF_3$	7140
Halon-1211	$CBrClF_2$	1890
Halon-2402	$CBrF_2CBrF_2$	1640
Carbon tetrachloride	CCl_4	1400
Methyl bromide	CH_3Br	5
Methyl chloroform	CH_3CCl_3	146
HCFC-21	$CHCl_2F$	151
HCFC-22	$CHClF_2$	1810
HCFC-123	$CHCl_2CF_3$	77
HCFC-124	$CHClFCF_3$	609

Continued

TABLE A.1 Global warming potentials (GWP) relative to carbon dioxide for the 100-year time horizon—cont'd

Industrial designation or common name	Chemical formula	GWP for 100-year time horizon (at date of publication)
HCFC-141b	CH_3CCl_2F	725
HCFC-142b	CH_3CClF_2	2310
HCFC-225ca	$CHCl_2CF_2CF_3$	122
HCFC-225cb	$CHClFCF_2CClF_2$	595
Hydrofluorocarbons		
HFC-23	CHF_3	14,800
HFC-32	CH_2F_2	675
HFC-41	CH_3F	92
HFC-125	CHF_2CF_3	3500
HFC-134	CHF_2CHF_2	1100
HFC-134a	CH_2FCF_3	1430
HFC-143	CH_2FCHF_2	353
HFC-143a	CH_3CF_3	4470
HFC-152	CH_2FCH_2F	53
HFC-152a	CH_3CHF_2	124
HFC-161	CH_3CH_2F	12
HFC-227ea	CF_3CHFCF_3	3220
HFC-236cb	$CH_2FCF_2CF_3$	1340
HFC-236ea	CHF_2CHFCF_3	1370
HFC-236fa	$CF_3CH_2CF_3$	9810
HFC-245ca	$CH_2FCF_2CHF_2$	693
HFC-245fa	$CHF_2CH_2CF_3$	1030
HFC-365mfc	$CH_3CF_2CH_2CF_3$	794
HFC-43-10mee	$CF_3CHFCHFCF_2CF_3$	1640
Perfluorinated compounds		
Sulfur hexafluoride	SF_6	22,800
Nitrogen trifluoride	NF_3	17,200
PFC-14	CF_4	7390

TABLE A.1 Global warming potentials (GWP) relative to carbon dioxide for the 100-year time horizon—cont'd

Industrial designation or common name	Chemical formula	GWP for 100-year time horizon (at date of publication)
PFC-116	C_2F_6	12,200
PFC-218	C_3F_8	8830
PFC-318	c-C_4F_8	10,300
PFC-3-1-10	C_4F_{10}	8860
PFC-4-1-12	C_5F_{12}	9160
PFC-5-1-14	C_6F_{14}	9300
PFC-9-1-18	$C_{10}F_{18}$	>7500
Trifluoromethyl sulfur pentafluoride	SF_5CF_3	17,700
Perfluorocyclopropane	c-C_3F_6	>17,340
Fluorinated ethers		
HFE-125	CHF_2OCF_3	14,900
HFE-134	CHF_2OCHF_2	6320
HFE-143a	CH_3OCF_3	756
HCFE-235da2	$CHF_2OCHClCF_3$	350
HFE-245cb2	$CH_3OCF_2CF_3$	708
HFE-245fa2	$CHF_2OCH_2CF_3$	659
HFE-254cb2	$CH_3OCF_2CHF_2$	359
HFE-347mcc3	$CH_3OCF_2CF_2CF_3$	575
HFE-347pcf2	$CHF_2CF_2OCH_2CF_3$	580
HFE-356pcc3	$CH_3OCF_2CF_2CHF_2$	110
HFE-449sl (HFE-7100)	$C_4F_9OCH_3$	297
HFE-569sf2 (HFE-7200)	$C_4F_9OC_2H_5$	59
HFE-43-10pccc124 (HGalden1040x)	$CHF_2OCF_2OC_2F_4OCHF_2$	1870
HFE-236ca12 (HG-10)	$CHF_2OCF_2OCHF_2$	2800
HFE-338pcc13 (HG-01)		
HFE-338pcc13 (HG-01)	$(CF_3)_2CHOH$	195
HFE-227ea	$CF_3CHFOCF_3$	1540

Continued

TABLE A.1 Global warming potentials (GWP) relative to carbon dioxide for the 100-year time horizon—cont'd

Industrial designation or common name	Chemical formula	GWP for 100-year time horizon (at date of publication)
HFE-236ea2	$CHF_2OCHFCF_3$	989
HFE-236fa	$CF_3CH_2OCF_3$	487
HFE-245fa1	$CHF_2CH_2OCF_3$	286
HFE-263fb2	$CF_3CH_2OCH_3$	11
HFE-329mcc2	$CHF_2CF_2OCF_2CF_3$	919
HFE-338mcf2	$CF_3CH_2OCF_2CF_3$	552
HFE-347mcf2	$CHF_2CH_2OCF_2CF_3$	374
HFE-356mec3	$CH_3OCF_2CHFCF_3$	101
HFE-356pcf2	$CHF_2CH_2OCF_2CHF_2$	265
HFE-356pcf3	$CHF_2OCH_2CF_2CHF_2$	502
HFE-365mcf3	$CF_3CF_2CH_2OCH_3$	11
Fluorinated ethers (continued)		
HFE-374pc2	$CHF_2CF_2OCH_2CH_3$	557
	—$(CF_2)_4CH(OH)$—	73
	$(CF_3)_2CHOCHF_2$	380
	$(CF_3)_2CHOCH_3$	27
Perfluoropolyethers		
PFPMIE	$CF_3OCF(CF_3)$ $CF_2OCF_2OCF_3$	10,300
Hydrocarbons and other compounds—Direct Effects		
Dimethylether	CH_3OCH_3	1
Chloroform	$CHCl_3$	31
Methylene chloride	CH_2Cl_2	8.7
Methyl chloride	CH_3Cl	13
	CH_2Br_2	1.54
Halon-1201	$CHBrF_2$	404
Trifluoroiodomethane	CF_3I	0.4

Annex 2: Table A.2.

TABLE A.2 Depletion of abiotic resources impact category and their respective characterization factors for the aggregation of single resources to resource depletion indicators for each impact category.

Resource	Unit	Characterization factor	Impact category	Unit of resource depletion indicator
Oil	kg OE[a] eq	42.86	Fossil fuels depletion	MJ
Natural gas	m^3	31.73		
Hard coal	kg	29.07		
Lignite	kg	8.50		
Phosphate rock	kg	0.25	Phosphate rock depletion	kg P_2O_5
Raw phosphate	kg	0.32		
Potash, potassium chloride	kg	0.10	Potash depletion	kg K_2O
Limestone/lime	kg	0.54	Lime depletion	kg CaO
Dolomite	kg	0.30		

[a]OE, *crude oil equivalent.*
Based on Brentrup, Küsters, Kuhlmann, and Lammel (2004).

Annex 3: Table A.3.

TABLE A.3 Land use impact category: Characterization factors for selected land use types.

Land use type	Unit	NDP[a]
Continuous urban area	ha/year	0.95
Industrial/commercial units	ha/year	0.95
Road/rail networks	ha/year	0.90
Discontinuous urban area	ha/year	0.85
Intensive arable land	ha/year	0.80
Extensive arable land	ha/year	0.70
Green urban area	ha/year	0.70
Intensive permanent pasture	ha/year	0.60
Extensive permanent pasture	ha/year	0.50

[a]NDP, *naturalness degradation potential.*

Annex 4: Table A.4.

TABLE A.4 "Human and eco-toxicity" impact category: Characterization factors (toxicity potentials) for cadmium (Cd) emissions to soil.

Category	Unit	Toxicity potential per kg Cd to soil
Human toxicity	DALY[a]	3.98×10^{-3}
Terrestrial ecotoxicity	kg 1,4-DCB-eq[b]	1.7×10^{2}
Aquatic ecotoxicity, fresh water	kg 1,4-DCB-eq	7.8×10^{2}
Aquatic ecotoxicity, marine	kg 1,4-DCB-eq	1.1×10^{5}
Sediment ecotoxicity, fresh water	kg 1,4-DCB-eq	2.0×10^{3}
Sediment ecotoxicity, marine	kg 1,4-DCB-eq	1.1×10^{5}

[a]DALY, *disability adjusted life-years (Goedkoop & Spriensma, 1999).*
[b]1,4-DCB-eq, *1,4-dichlorobenzene-equivalents (Huijbregts, 2001).*

Annex 5: Table A.5.

TABLE A.5 Acidification and terrestrial eutrophication impact categories: Regionalized characterization factors (acidification potentials for SO_2, NO_x, NH_3 and terrestrial eutrophication potentials for NO_x, NH_3).

	Acidification potential (in kg SO_2-equivalents per kg emission)			Terrestrial eutrophication potential (in kg NO_x equivalents per kg emission)	
Emission source region	SO_2	NO_x	NH_3	NO_x	NH_3
Switzerland	1.00	0.28	1.30	1.00	5.00
Austria	1.00	0.27	1.30	0.89	4.20
Belgium	1.00	0.49	1.00	1.20	2.90
Denmark	1.80	0.88	1.50	1.60	2.50
Finland	5.00	1.90	6.40	3.50	11.50
France	1.10	0.43	2.00	1.30	6.40
Germany	1.30	0.53	1.50	1.50	4.60
Greece	0.066	0.037	0.13	0.27	1.50
Ireland	0.57	0.34	0.79	0.52	1.00
Italy	0.46	0.13	0.59	0.60	2.80
Luxembourg	1.30	0.50	1.50	1.40	4.40
Netherlands	0.92	0.51	1.00	1.10	2.30
Norway	3.80	1.20	6.00	1.60	6.20
Portugal	0.18	0.08	0.28	0.49	2.40
Spain	0.22	0.10	0.27	0.52	2.00
Sweden	3.80	1.30	4.40	2.10	5.70
United Kingdom	0.86	0.43	1.50	0.76	1.70
Western Europe, average	0.79	0.41	1.30	0.99	3.70
Eastern Europe, average	1.60	0.70	1.80	1.70	5.00
Europe, average	1.20	0.50	1.60	1.20	4.30

Modified from Huijbregts (2001).

Annex 6: Table A.6.

TABLE A.6 Regionalized fate factors for airborne NO_x and NH_3 emissions to approximate the related emissions to marine surface waters for the "Aquatic eutrophication" impact subcategory.

Emission source region	Fate factors	
	NO_x	*NH_3*
Switzerland	0.094	0.051
Austria	0.088	0.048
Belgium	0.240	0.230
Denmark	0.290	0.430
Finland	0.200	0.250
France	0.230	0.250
Germany	0.170	0.140
Greece	0.180	0.230
Ireland	0.470	0.460
Italy	0.190	0.210
Luxembourg	0.170	0.110
Netherlands	0.280	0.260
Norway	0.280	0.450
Portugal	0.160	0.230
Spain	0.170	0.160
Sweden	0.240	0.330
United Kingdom	0.390	0.430
Western Europe, average	0.250	0.240
Eastern Europe, average	0.089	0.072
Europe, average	0.210	0.160

Modified from Huijbregts and Seppälä (2000).

Annex 7: Table A.7.

TABLE A.7 POCP values for individual volatile organic compounds.

Compound	POCP factor (g ethylene/100 g gas)	Emission factor reference
Acid		
Acetic acid	9.7	1
Alcohol		
1-Butanol	61.2	1
1-Butanol, 2-methyl-	40.7	1
1-Butanol, 3-methyl-	41.2	1
Cyclohexanol	44.6	1
1-Hexanol, 2-ethyl	38.3	3
1-Pentanol	38.3	3
1-Propanol	54.3	1
1-Propanol, 2-methyl-	37.5	1
2-Butanol, (*R*)-	38.3	3
2-Propanol, 2-methyl-	12.3	1
Ethanol	38.6	1
Isopropyl Alcohol	14.0	1
Aldehyde		
Acetaldehyde	64.1	1
Butanal, 3-methyl-	79.5	1
Methacrolein	92.0	2
Propanal, 2-methyl-	51.4	1
Alkane		
Butane	35.2	1
Butane, 2-methyl-	40.5	1
Cyclohexane	29.0	1
Cyclohexane, methyl-	65.0	2
Ethane, 1,1,1-trichloro-	0.9	4
Heptane	49.4	1

Continued

TABLE A.7 POCP values for individual volatile organic compounds—cont'd

Compound	POCP factor (g ethylene/100 g gas)	Emission factor reference
Heptane, 2-methyl-	34.0	2
Hexane, 2-methyl-	41.1	1
Hexane, 3-methyl-	36.4	1
n-Decane	38.4	1
n-Dodecane	35.7	1
Nonane	41.4	1
Nonane, 2-methyl-	35.0	2
Nonane, 3-methyl-	39.0	2
Nonane, 4-methyl-	35.0	2
Octane	45.3	1
Octane, 2-methyl-	34.0	2
Pentane	39.5	1
Pentane, 2,2-dimethyl-	43.0	3
Pentane, 2-methyl-	42.0	1
Pentane, 3-methyl-	47.9	1
Propane	17.6	1
n-Hexane	48.2	1
Tetradecane	46.0	2
Undecane, 2,2-dimethyl-	61.6	2
Alkene		
2 Butene	114.6	1
1 Pentene	99.7	1
Aliphatic Hydrocarbon		
Methylene chloride	6.8	1
Tetrachloroethylene	2.9	1
Trichloroethylene	32.5	1
Trichloromethane	2.3	1
Trichloromonofluoromethane	2.5	3

TABLE A.7 POCP values for individual volatile organic compounds—cont'd

Compound	POCP factor (g ethylene/100 g gas)	Emission factor reference
Aromatic Hydrocarbon		
Benzene	21.8	1
Benzene, 1,4-dichloro-	90.0	3
Benzene, 1-ethyl-2-methyl-	132.0	1
Benzene, propyl-	63.6	1
Ethylbenzene	73.0	1
Mesitylene	138.5	4
o-Xylene	105.3	1
p-Xylene	101.0	1
Styrene	14.2	1
Toluene	63.7	1
Ester		
Acetic acid, butyl ester	24.1	1
Acetic acid, methyl ester	4.6	1
Butanoic acid, ethyl ester	17.0	3
Butanoic acid, methyl ester	17.0	3
Ethyl Acetate	21.3	1
Heptanoic acid, ethyl ester	17.0	3
Hexanoic acid, 2-methylpropyl ester	17.0	3
Hexanoic acid, butyl ester	17.0	3
Hexanoic acid, ethyl ester	17.0	3
Hexanoic acid, methyl ester	17.0	3
Hexanoic acid, propyl ester	17.0	3
Isopentyl hexanoate	17.0	3
Methyl valerate	17.0	3
n-Propyl acetate	29.0	1
Octanoic acid, ethyl ester	17.0	3
Pentanoic acid, ethyl ester	17.0	3

Continued

TABLE A.7 POCP values for individual volatile organic compounds—cont'd

Compound	POCP factor (g ethylene/100g gas)	Emission factor reference
Propane, 2-ethoxy-2-methyl-	17.0	3
Propanoic acid, 2-methyl-, methyl ester	17.0	3
Ether		
Dymethyl ether	17.4	1
Methylal	55.5	4
Tetrahydrofuran	0.0	3
Ketone		
2,3-Butanedione	42.0	3
2-Butanone	37.3	1
2-Heptanone	42.0	3
2-Hexanone	57.2	1
2-Pentanone	54.8	1
Acetone	9.4	1
Acetophenone	42.0	3
Cyclohexanone	29.9	1
Methyl Isobutyl Ketone	49.0	1
Sulfide		
Carbon disulfide	4.8	4
Disulfide, dimethyl	4.8	4
Terpene		
t-Terpinene	57.3	3
a-Pinene	68.0	2
ß-Pinene	33.0	2
D-Limonene	71.0	2

1—Derwent, R. G., Jenkin, M. E., Saunders, S. M., Pilling, M. J. (1998). Photochemical ozone creation potentials for organic compounds in north west Europe calculated with a master chemical mechanism. *Atmospheric Environment, 32,* 2429–2441.
2—Derwent, R. G., Jenkin, M. E., Passant, N. R., Pilling, M. J. (2007). Reactivity-based strategies for photochemical ozone control in Europe. *Environmental Science and Policy, 10,* 445–453.
3—POCP factor was calculated as average of the families determined in literature.
4—Altenstedt, J., Pleijel, K. (1998). *POCP for individual VOC under European conditions.* IVL Rapport B-1305. IVL, Publikationsservice, Box 21060, S-100 31 Stockholm.

References

Brentrup, F., Küsters, J., Kuhlmann, H., & Lammel, J. (2004). Environmental impact assessment of agricultural production systems using the life cycle assessment methodology: I. Theoretical concept of a LCA method tailored to crop production. *European Journal of Agronomy, 20*(3), 247–264.

Goedkoop, M. S. R. T., & Spriensma, R. (1999). *The Eco-indicator 99: A damage oriented method for life cycle impact assessment methodology* (3rd ed., pp. 1–88). Amersfoort, Plotterweg: PRé Consultants.

Huijbregts, M. (2001). Uncertainty and variability in environmental life-cycle assessment. *The International Journal of Life Cycle Assessment, 7*, 173.

Huijbregts, M. A., & Seppälä, J. (2000). Towards region-specific, European fate factors for airborne nitrogen compounds causing aquatic eutrophication. *The International Journal of Life Cycle Assessment, 5*(2), 65–67.

Index

Note: Page numbers followed by *f* indicate figures, *t* indicate tables, and *b* indicate boxes.

A

Acidification
  calculation exemplification, 39–40*b*
  category outline, 39
  SO_2 equivalence factors, 38*t*
Acute toxicity tests, 35–36
Agricultural systems
  allocation, 122–125
  fertilization, 133–137
  functional unit, 118–119
  inputs, 118
  land use, 128–133
  life cycle inventory (LCI)
    quantification bottlenecks, 125–127
  parameters for, 137–139
  system boundaries, 119–122
Allocation, 14
  agricultural systems, 122–125
Analysis tool, 3–4*f*
AoPs. *See* Areas of protections (AoPs)
Aquatic ecosystems, 38
Arbitrary level, 75
Areas of protections (AoPs)
  endpoint indicators, 67*t*
  environmental impacts, 66*f*

B

Beyond sustainability, 180–182

C

Carbon certifications, 102–104
Carbon credits, 106–108
  policy strategies, 106–111
Carbon ecosystem, 138
Carbon footprint
  carbon certifications, 102–104
  carbon credits, 106–108
  carbon market trades, 111–115
  concepts, 92*t*
  downstream carbon footprint, 101–102
  market carbon, 104–106
  onsite carbon footprint, 100
  policy strategies, 108–111
  standardization, 91–102
    evaluation of, 95–96
    greenhouse gases, 94–95
    methodological applications, 96–97
    system delimitations, 98–100
  upstream carbon footprints, 100–101
Carbon market trades, 111–115
Carbon quantification, 91
CBA. *See* Cost-benefit analyzes (CBA)
Characterization factor, 31*t*
Chronic toxicity tests, 36
Comparability assessment, 11
Cost-benefit analyzes (CBA), 152
Crop farms, 133–137

D

Data interpretation
  evaluation, 23
  identification, significant issues, 22
Data validation, 14
Decision-making, 87
Depletion global, 43–51
Direct land use changes (dLUC), 56–57
Downstream carbon footprint, 101–102
Downstream truncation, 15

E

Eco-efficiency, 160–162
Ecological crisis, 175
Economic dimension, 175–176
Economic indicators, 171
Ecotoxicity
  calculation exemplification, 37*b*
  category outline, 37
Efficiency of energy renewability (ERenEf), 46
Emissions trades, 113
Endpoints, 30*f*
Energy assessments, 43–51
  calculation exemplification, 49–52*b*
  category outline, 49, 49*t*
  energy efficiency indicators, 48*t*
Energy efficiency indicators, 48*t*
Energy return on energy invested (EROI), 45
Environmental and resource profile analysis
    (REPA), 1

Environmental awareness, 171
Environmental dimension, 175
Environmental effects, 66
Environmental impacts, 14
ERenEf. *See* Efficiency of energy renewability (ERenEf)
EROI. *See* Energy return on energy invested (EROI)
Eutrophication potential (EP)
  calculation exemplification, 42*b*
  category outline, 41
  PO_4 equivalence factors, 41*t*
Evaluation process, 89–90

F

Fertilization, agricultural crops, 133–137
Fossil energy ratio (FER), 47
Freshwater aquatic ecotoxicity potential (FAEP), 36
Functional unit approach, 118–119

G

Global warming
  calculation exemplification, 62*b*
  category outline, 62*t*
  characterization factors, 61*t*
Greenhouse gases, 94–95
Grouping, 71

H

Human toxicity potential (HTP)
  areas of protection (AoPs), 63–66
  category outline, 65*t*
  substances and bottlenecks, 64*t*
Hybrid method, 15, 16*t*
Hydrogen sulfide (H_2S), 39–40

I

iLUC. *See* Indirect changes in the use of land ground (iLUC)
Impact categories
  acidification potential, 38–39
  consensus at international level, 30*t*
  ecotoxicity, 35–37
  energy assessments, 43–51
  eutrophication potential (EP), 40–42
  global warming potential, 59–62
  human toxicity potential (HTP), 63
  land use, 56–58
  midpoints and endpoints model, 30*f*
  ozone depletion potential (stratospheric) (ODP), 31–32
  photochemical ozone creation potential (tropospheric) (POCP), 33–34
  water footprint, 52–55
Indirect changes in the use of land ground (iLUC), 56–57
Input/output method, 15, 16*t*
Institutional sustainability, economic aspects
  manifestation of, 148–151
  shared economy, 151–155
  sustainable added value, 155–169
International Organization for Standardization (ISO), 7
Inventory assessment, 13*f*
Inventory stage, 125–127

L

Land use
  calculation exemplification, 59*b*
  category outline, 58*t*
  direct and indirect land use, 56*f*
  ecosystem quality models, 131*t*
  ecosystems, 58*t*
  environmental impacts, 132*t*
  natural capital index (NCI), 130*f*
Lateral truncation, 15–17
LCA. *See* Life cycle assessment (LCA)
LCC. *See* Life cycle cost (LCC)
LCEA. *See* Life cycle energy analysis tool (LCEA)
LCEE. *See* Life cycle energy efficiency (LCEE)
LCI. *See* Life cycle inventory (LCI)
LCIA. *See* Life cycle impact assessment (LCIA)
LCSA. *See* Life cycle sustainability assessment (LCSA)
LCT. *See* Life cycle thinking (LCT)
Life cycle assessment (LCA), 2
  application of, 4*f*
  normative evolution, 4*f*
  phases
    boundaries, 23–26
    interpretation, 22–23
    life cycle impact assessment (LCIA), 18–21
    life cycle inventory (LCI), 13–17
    objective and scope, 8–12
  standard methodology, 7
  timeline of, 3*f*
Life cycle cost (LCC)
  internal and external costs, 84*f*
  potential categories, 85*t*
Life cycle energy analysis tool (LCEA), 43
Life cycle energy efficiency (LCEE), 47

Life cycle impact assessment (LCIA)
implementation of, 18
weighting methods, 20–22*t*
Life cycle inventory (LCI), 13–17
advantages and bottlenecks, 16*t*
agricultural systems
quantification bottlenecks, 125–127
allocation, 14
databases, 17*t*
data collection, 14
data validation, 14
downstream truncation, 15
hybrid method, 15
input/output method, 15
lateral truncation, 15–17
process-based methods, 14–15
software, 17*t*
upstream truncation, 15
Life cycle sustainability assessment (LCSA)
decision-making problem, 87
evaluation matrix, 89–90
sensitivity analysis, 90
structure for, 86*f*
Life cycle thinking (LCT), 77–79

M

Marine aquatic ecotoxicity potential (MAEP), 36
Market carbon, 104–106
policy strategies, 104–106, 108–111
Methodology, 74
Midpoints, 30*f*
Multicriteria analysis (MCA), 86–87

N

Natural capital index (NCI), 130*f*
Net energy balance (NEB), 44
Net energy gain (NEG), 44
Net energy production (NEP), 44
Net energy ratio (NER), 45
Net energy value (NEV), 44–45
Net energy yield (NEY), 44
Normalization, 71–73
Normative evolution, 4*f*

O

ODP. *See* Ozone depletion potential (ODP)
Onsite carbon footprint, 100
Opportunity costs
benchmark, 165*f*
exemplification, 164*t*
Ozone depletion potential (ODP)
calculation exemplification, 32–33*b*
category outline, 32*t*
CFC-11 equivalence factors, 31*t*

P

PEBs. *See* Primary energy balances (PEBs)
Photochemical ozone creation potential (POCP), 34–35*b*
calculation exemplification, 34–35*b*
category outline, 34*t*
Primary energy balances (PEBs), 46
Process-based methods, 14–15, 16*t*
Process mapping, 11–12

Q

Qualitative process, 74–75
Quantification bottlenecks, 125–127

R

Reference system, 11
Renewable factor (RF), 46

S

SALCA. *See* Swiss Agricultural Life Cycle Assessment (SALCA)
Semiquantitative process, 74–75
Sensitivity analysis, 90
Shared economy, 151–155
Social dimension, 176–177
Social life cycle assessment (S-LCA)
indices, subindices, and indicators of, 81–83*t*
stakeholder categories, 79*f*, 80*t*
Social metrics, 177–180
Soil protection, 136
Stratospheric ozone. *See* Ozone depletion potential (ODP)
Sustainability
economic dimension, 175–176
environmental dimension, 175
indexes, 75
metrics and indicators, 86, 177–180
propagation of, 180
social dimension, 176–177
Sustainable added value
absolute measures, 155–160
relative measures, 156–160
Sustainable evaluation, 2
Swiss Agricultural Life Cycle Assessment (SALCA), 135
System boundaries
agricultural systems, 119–122
life cycle assessment

System boundaries *(Continued)*
cradle to cradle, 24
cradle to gate, 24
cradle to grave, 24
gate to the gate, 24–25
well to wheel, 26

T

TEP. *See* Toxic equivalency potential (TEP)
Terrestrial ecosystems, 38
Terrestrial ecotoxicity potential (TETP), 36
Theoretical approach, 77
Timeline, 3*f*
Toxic equivalency potential (TEP), 63
Toxicity potential (ETP), 35
Tropospheric ozone. *See* Photochemical ozone creation potential (POCP)
Truncation
downstream, 15
lateral, 15–17
upstream, 15

U

Upstream carbon footprints, 100–101
Upstream truncation, 15

V

Volatile organic compounds (VOCs), 34–35*b*

W

Water footprint
calculation exemplification, 55*b*
category outline, 54*t*
criteria and divisions, 53*f*
Weighting, 20–22*t*, 74–75

Made in the USA
Coppell, TX
31 March 2026

75121396R00125